sun protection

Marie Borrel

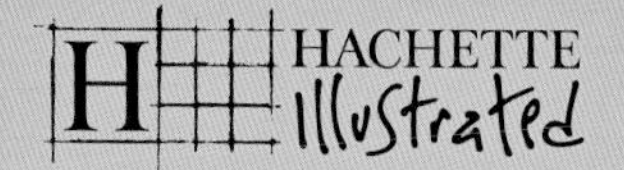

contents

introduction 4
how to use this book 7
how do you treat the sun and how does it treat you? 8

useful addresses 124
index 125
acknowledgements 126

01 how good are your resistance levels? 12
02 moisturize your skin 14
03 body and facial care 16
04 water, water and more water... 17
05 eat more carrots! 18
06 get plenty of 'good' fats 20
07 don't forget vitamin E 22
08 tomatoes have a lot to offer 24
09 you can rely on this trio! 26
10 selenium and zinc, a high-power duo 28
11 try 'pre-sun' tablets 30
12 take a course of horsetail 32
13 diet with caution 33
14 a nice cup of tea 34
15 sunbathing and skin problems 36
16 avoid the sun if you have acne 38
17 beware of sun blisters 40
18 discourage those brown blotches 42
19 steer clear of bergamot 44
20 protect your scars 46
case study 47

21 enjoy the benefits of the sun 50
22 when the weather gets warmer… 52
23 avoid the hottest part of the day 54
24 screen those dangerous UV rays 56
25 a few facts about sun creams 58
26 choose the right factor 60
27 maximum protection: total sun block 62
28 self-tanning creams: pros and cons 64
29 regular and generous applications 66
30 slop it on all over 67
31 think carefully before using a sunbed 68
32 relief from psoriasis 70
33 look after your eyes 72
34 hats and sunglasses for the kids 74
35 beware of the mirror effect 76
36 just how dangerous is the snow? 77
37 use aloe vera gel 78
38 don't forget shea-butter 80
39 don't be fooled by clouds 82
40 be extra careful with babies 84
case study 85

41 take a shower 88
42 use after-sun lotion 90
43 St John's wort oil, a natural balm 92
44 make a fuss of your hair 94
45 what seaweed can do for you 96
46 give your skin a good night's sleep 98
47 scrub your body 99
48 soothing the effects of sunburn 100
49 try healing plants 102
50 dealing with heatstroke 104
51 act quickly to combat sunstroke 106
52 keep an eye on your legs 108
53 soothing swollen legs 110
54 getting rid of excess fluid 112
55 tackling perspiration 113
56 finding the right points 114
57 treat yourself to a beauty massage 116
58 dealing with sun allergies 118
59 sun allergies and homeopathy 120
60 dare to be pale 122
case study 123

introduction

sunbathing without burning

Our relationship with the sun is vital and profound. We wouldn't survive without it. Its heat and light are absolutely indispensable. In Nordic countries, many people suffer from a type of depression caused by the lack of sunlight during the winter months. So the sun even has an enormous effect on our moods. It is also involved in the synthesis of vitamin D, which strengthens our bones. And it orchestrates our biological clock, as it is responsible for the basic rhythms of day and night, winter and summer. It's hardly surprising that ancient civilizations, from the Egyptians to the Incas, worshipped the sun as a god and the original source of all life.

Those Victorian wide-brimmed hats

However, the relationship between people and the sun has varied a great deal over time. In the past, fashionable ladies hid their faces under wide-brimmed hats to preserve a milky-white complexion and inspected themselves for the slightest trace of redness. In those days, a tan indicated low social status: only labourers and farm workers caught the sun.

Fashions have changed a lot since then. Nowadays, a well-bronzed face in the middle of winter lets everyone know you've been lucky enough to take a holiday in some exotic location. A suntan has become not merely a sign of health, but of success and wealth. As a result, everyone wants one.

Unfortunately, though, we're not all equally equipped to enjoy the sunshine. Some people tan quickly and safely, while others are soon as red as Rudolph's nose! Between these two extremes lies a range of possibilities. How we react depends on our genetic make-up, and there's nothing we can do to change that.

Melanocytes and melanin

Our skin changes colour in the sunshine because we have cells called melanocytes, which produce the pigment melanin. Its purpose is to colour the skin to protect it from the sun's rays. A problem arises, however, because some people have fewer and less active melanocytes than others. What's more, the melanin itself isn't always identical. This pigment is sometimes black (eumelanin) and sometimes red (pheomelanin). Our melanocytes secrete a mixture of the two, in different proportions according to each individual. Eumelanin is a thousand times more effective than pheomelanin in protecting the skin. In addition, the latter can cause cell mutations when secreted in excessive amounts. Production of eumelanin is regulated by the MC1R gene, and people who have

inherited mutations in this gene tend to produce more pheomelanin, which is less effective in protecting against ultraviolet light. As a result, the skin is more prone to cancer. Those who secrete mostly black melanin are therefore at a big advantage when compared with those who produce mainly red.
Some people think it's fine to get a suntan as quickly as they can, since they've got melanin to protect them from the sun's rays. However, even when deeply tanned, the skin is still harmed by those ultraviolet rays. A tan protects us from sunburn, but it doesn't prevent some rays from penetrating to the very heart of the skin's cells, where they may cause serious damage.

Some damage can't be repaired

Fortunately, we possess our own internal repair mechanism: when a cell is harmed (and, specifically, when its DNA is damaged), enzymes set to work and try to mend it. In most cases they succeed in repairing the damage, but occasionally there's nothing they can do. In those cases, the defective cell is eliminated from the body. However, if we expose ourselves to the sun too often and for too long, our repair mechanism cannot cope and damaged DNA accumulates. If too much DNA and too many cells are damaged, the body is unable to eliminate all of them, thus increasing the risk of skin cancer.
To enjoy the benefits of the sun without too much harm, we need to limit the effect if its power. We can do this by nourishing and caring for the body in different ways. It is also important to give our skin time to adapt to the sun, so use the appropriate protection products and moisturize regularly so that the body is better able to resist damage.
Apart from this general advice, which forms the basis of sensible sunbathing, certain skin types require special advice. So try to work out which skin type you are, and you'll be able to enjoy the innumerable benefits of the sun without suffering many of the disadvantages.

how to use this book

FOR YOUR GUIDANCE

> A symbol at the bottom of each page will help you to identify the natural solutions available:

:

Herbal medicine, aromatherapy, homeopathy, Dr Bach's flower remedies – how natural medicine can help.

Simple exercises – preventing problems by strengthening your body.

Massage and manipulation – how they help to promote well-being.

Healthy eating – all you need to know about the contribution it makes.

Practical tips for your daily life – so that you can prevent instead of having to cure.

Psychology, relaxation, Zen – advice to help you be at peace with yourself and regain serenity.

> A complete programme that will solve all your health problems. Try it!

This book provides advice to enable you to enjoy the sun while minimizing the risks. It's divided into four sections:

- **A questionnaire** to help you assess what mistakes you might be making.
- **The first 20 tips** prepare you for going out in the sun: food, plants, precautions and special cases.
- **The next 20 tips** tell you how best to protect your skin when you're in the sun, whether at the seaside, in the country or in the mountains.
- **The final 20 tips** deal with how to maintain your tan once you have returned from your holidays and how to treat sunburn and other problems.

Case studies at the end of each section of tips relate different people's sun-bathing experiences.

You can either work methodically through the book starting at Tip 1 and ending at Tip 60, putting each piece of advice into practice, or, if you prefer, you can pick out those recommendations that seem particularly effective or that best suit your habits and lifestyle.
It's up to you!

how do you treat the sun and how does it treat you?

Reply honestly to the statements below.

true	false	1. I never use any sun cream or protection.
true	false	2. The sun makes me break out in spots.
true	false	3. My legs swell when I'm hot.
true	false	4. I hate wearing sunglasses and sunhats.
true	false	5. I always go on a diet before going on holiday.
true	false	6. I sweat a lot.
true	false	7. I don't eat a lot of fruit and vegetables.
true	false	8. I'm allergic to the sun.
true	false	9. I love roasting in the midday sun.

If you have answered TRUE to statements 2, 5 and 7, then Tips **1** to **20** are the most suitable for you.

If you have answered TRUE to statements 1, 4 and 9, then go straight to Tips **21** to **40**.

If you have answered TRUE to statements 3, 6 and 8, then read Tips **41** to **60** first.

›› The sun doesn't have to be a danger to your skin, but it can easily become one. **The skin can protect itself against the sun provided you are sensible and take your time.** The best preparation is year-round skincare.

›››› **Diet is particularly important:** feed your skin cells with vital vitamins, minerals and essential fatty acids.

›››››› Make sure you hydrate both inside and out by moisturizing your skin and **drinking plenty of fluids**.

›››››››› Finally, try to remember that **eating greens is good for your skin.**

20
TIPS

01 how good are your resistance levels?

We don't all tan as easily as one another. It's not fair, but that's the way it is. So it's important to know your skin type and to adapt your approach accordingly – especially as your reserves of resistance are depleted over the years, particularly if you do a lot of sunbathing.

Blond, brunette, redhead …

Different skin types react to the sun in different ways. Some people turn as red as lobsters almost as soon as they are exposed to the sun, while others quickly achieve a lovely golden colour. No sun cream can alter your basic potential: it would take a magic wand to do that. Everybody's skin corresponds more or less to a particular type. These categories involve such factors as the colour

DID YOU KNOW?

> There are five general skin types:

– Very pale complexion, red hair, freckles, light eyes: the skin burns but doesn't tan.

– Pale complexion, blond hair, light eyes: the skin burns but eventually becomes slightly tanned.

– Less pale complexion, brown hair, light eyes: sunburn frequently occurs, but it's possible to achieve a tan.

– Dark complexion, brown hair, brown eyes: rarely suffers from sunburn.

– Very dark complexion, black hair, dark eyes: tans rapidly.

of your skin, hair and eyes. To these are added individual genetic features, which mean, for instance, that even within a certain colour range, skins can have a varying tolerance.

Reserves at risk

Furthermore, we all possess reserves of resistance that we draw upon every time we expose ourselves to the sun. The extent of these reserves varies from individual to individual and depends on our capacity to produce melanin (which colours the skin to make it more difficult for rays to penetrate) and keratin (which protects the skin by thickening it). Both substances provide a defence against oxidation, which accelerates ageing. The more you turn to your reserves, the more you will deplete them, and the day will eventually come when there's nothing left at all: the skin will produce little or no melanin and the keratin will be of such poor quality that you will be at high risk from oxidation. At this point, the sun becomes very dangerous.

Take a good look at your skin and delve back into the past to do a stocktake of your reserves of resistance. If you've been exposed to the sun a great deal since you were a child, be extra careful, because there's a chance that your reserves are running low.

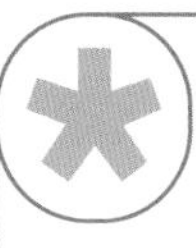

KEY FACTS

* Different skin types react differently to the sun. We should adapt our approach according to our skin type.

* We have a capacity to resist the sun, but this diminishes with the passage of time.

* The more sunbathing we do, the more careful we need to be.

Skin in good condition is better able to tolerate the sun. Moisturize it throughout the year and it will thank you when summer comes. Regularly use moisturizing creams and also treat yourself to skin scrubs and exfoliants to enable the active particles in the creams to penetrate fully.

02 moisturize your skin

Moisturizing cosmetics and plants

A moisturized skin has much better resistance to the sun's rays. Heat, salty air at the seaside, chlorine in swimming pools and the wind at high altitudes all have an effect of drying the skin. And the more it dries out, the more the sun may harm it. So it's best to get a head start and moisturize all year round.

●●● DID YOU KNOW?

> Dry skin obviously lacks moisture but, contrary to what you might think, moist, greasy skin can be dehydrated, too.

> Think about it: if excess sebum blocks the pores, the skin will not be hydrated.

> Direct contact with water tends to dry the skin, as it affects the invisible layer of sebum that limits water loss.

> After you've had a bath or removed make-up with soap and water, use a cream to restore the moisture.

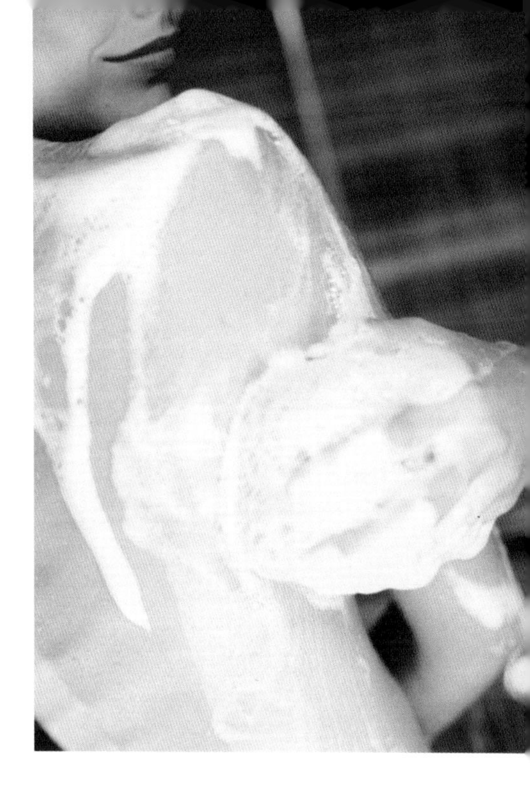

Cosmetics designed for this purpose are usually very effective. Some natural products, like aloe gel and shea-butter, are also very useful and you can apply them as often as you like.

Rapid removals

Remember to accompany this regime with frequent but gentle skin scrubs. Your skin breathes and plays its part in removing waste materials from the body, but in doing so it causes a problem: it eliminates its dead cells by bringing them to the surface, thus clogging its upper layer. A gentle scrub (using exfoliating cream or gel) accelerates the removal process and safely rids the skin of these harmful dead cells. If you have moist, greasy skin, you can do this once a week, but dry skin, being more sensitive, will tolerate this treatment only once a fortnight. If you eradicate the skin's dead cells, your moisturizing creams will be able to penetrate more rapidly and deeply.

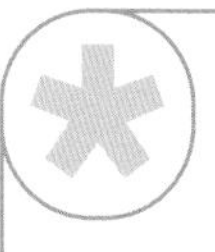

KEY FACTS

* Skin that has been well moisturized both morning and evening is better protected against the sun.

* Regularly use gentle skin scrubs to remove dead cells, help the skin breathe and improve the penetration of moisturizers.

03 body and facial care

Your face sees the light of day more than your body and becomes more acclimatized to the sun when fine weather comes. Both your face and your body need to be treated with care ... but not in the same way.

Morning and evening Facial skin is exposed all year round. It is subjected to the cold of winter, the dampness of autumn and the fierce heat of the summer sun. It therefore needs moisturizing morning and evening. In winter, one session a week of a skin scrub and moisturizing milk is adequate for the body.

Every other day In summer, however, you need to take much more care of your body's skin. Facial skin gradually adapts to the sun's rays and there are plenty of day creams containing filters to protect it from UV light. When you go for your first sunbathing session, it will already have produced a little melanin. The body's skin, however, with the exception of the arms and legs, will often not have seen any sunlight for a long time. So make sure you protect it well by moisturizing it more frequently (at least once every other day and certainly after every sunbathing session).

DID YOU KNOW?

> Remember to moisturize your lips as they are particularly sensitive. The skin here is very fine with many blood vessels just below the surface. In winter, lips often suffer from chapping, while in summer they dry and crack in the sun. Use a lip balm regularly, both in summer and winter.

KEY FACTS

* Your face needs daily skincare all year round.

* The body is always covered in winter, so it needs less regular care then, but you must step up your care regime in summer.

04 water, water and more water...

Your skin is also 'moisturized' from within. Drink plenty of water each day: your body needs it for a multitude of reasons, skin protection being just one of them.

A vital liquid Your cells bathe in an interstitial liquid, from which they draw their nourishment and into which they discharge toxins. When this liquid contains too much waste, the cells can no longer be adequately nourished. It therefore needs regular renewal. The answer is simple: drink more water.

Don't wait until you're thirsty Doctors recommend that you drink, on average, one and a half litres (three pints) of water per day. But some metabolisms are very economical with water, while others need more of it. One thing is certain: you need to drink before you feel thirsty, preferably small, regular amounts (a glass at a time). It's best to drink mineral water, which will provide the minerals your skin needs: selenium and zinc (see Tip 10), for example, but also calcium and magnesium to keep the skin soft, and potassium to stop it drying.

DID YOU KNOW?

> Sunbathing means heat and therefore perspiration.

> Although sweating is a healthy and necessary response, it causes the skin to rapidly lose water, so make sure you have a bottle of water with you when you are lying in the sun.

KEY FACTS

* Skin cells are surrounded by liquid, which you need to renew regularly.

* Drink before you feel thirsty, often and in small amounts (a glassful at a time).

* Take a bottle of water with you when you go to the beach.

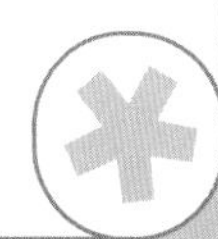

It's not by chance that beta-carotene is the main ingredient in food supplements recommended for sunbathers. This substance, which the body converts into vitamin A, is beneficial in many ways. For instance, it protects us from free radicals and improves the skin's ability to renew itself.

05 eat more carrots!

Beware of singlet oxygen!

The sun's ultraviolet rays cause oxidation in the body and thus greatly increase the number of free radicals, which are responsible for ageing and contribute to some kinds of skin cancer. Singlet oxygen is a prime example of how destructive free radicals can be: it dries the skin, harms the cell walls, damages the connective tissue and collagen, and even creates molecules capable of

DID YOU KNOW?

> We should consume 30 to 50 mg of beta-carotene each day. As there are 8 mg per 100 g (4 oz) of carrots, you need to eat 500 g (1 lb) of carrots every day if that is your only source of vitamin A.

> However, it's easy to avoid such a monotonous diet. For more variety, try apricots (1.5 mg per 100 g), spinach (4 mg), fresh mangoes (3 mg), red pepper (3.5 mg) and cabbage (3 mg).

> Unlike most vitamins, beta-carotene is absorbed better when carrots are cooked. Steam them with a sprinkling of unrefined oil.

suppressing the body's immune system. Before exposing the skin to the sun, you need to provide it with protection against these damaging effects.
Nature offers the ideal protection: beta-carotene. This nutrient provides us with a defence against free radicals and stimulates the skin's capacity to repair itself more quickly and effectively after the damage done by the sun's rays.

Apricots, melons, spinach, tomatoes...

As its name indicates, beta-carotene is found in large quantities in carrots. It is a vitamin A precursor: in other words, the cells lining the intestines split beta-carotene to form vitamin A. Just enough of the vitamin is produced to meet the body's needs and no more, so there's no risk of overdosing: you can eat as many carrots as you like. The only side-effect that comes when you eat them in vast quantities is that they can give a yellow tinge to the skin.
Beta-carotene is also found in all orange-coloured fruits and vegetables, such as apricots, pumpkins and melons, as well as in tomatoes and some green vegetables, like cabbages, spinach and broccoli.

KEY FACTS

* Solar rays cause an increase in free radicals, which accelerate ageing and can cause cancer.

* Beta-carotene helps protect us from the damaging effects of free radicals.

* We should consume between 30 and 50 mg of it each day.

06 get plenty of 'good' fats

Like all the other cells in our body, those that make up the dermis and epidermis are mainly composed of fats. To help them avoid being harmed by the sun's rays, it's vital that they have a year-round supply of good-quality fats. So make a bee-line for unrefined oils.

DID YOU KNOW?

> Unlike red meat, oily fish contain 'good' fatty acids. Try to eat salmon, mackerel and sardines regularly.

> Duck is the exception to the animal-fat rule: it also contains 'good' fatty acids.

End the vicious circle

When cell walls lack fatty acids, they stiffen. Cellular exchanges deteriorate, and the entire process by which the cells are nourished is suppressed. When skin tissues suffer such a shortage, the effects are immediately visible: the skin becomes flaccid and soft; wrinkles start to appear. More importantly, the skin's resistance to sunshine is reduced. What's more, free radicals, created by exposure to the sun, damage the cell walls and oxidize the fats contained in them. However, certain essential fatty acids, when consumed regularly, can put an end to this vicious circle.

The Omega Factor

Not all fatty acids are the same. Animal fats (except those in oily fish) contain mainly saturated fats, which are no help to the cell walls and, in addition, help to block the body's arteries. Vegetable fats, on the other hand, contain unsaturated fats. These are called 'essential' because they are indispensable to the body.

There are two groups of essential fatty acids: omega 3s and omega 6s. Together they help keep the cell walls in good condition and regenerate fats oxidized by solar rays. So eat cold-pressed, unrefined oils (fatty acids are damaged by heat) and make sure you consume a variety of them (each has its own particular virtues). Try walnut oil (delicious with green salads), sesame oil (with its delicate, grilled-hazelnut flavour) and others such as rape seed, sunflower, grape seed and, of course, olive oil.

> **Evening-primrose and borage oils, which are rich in both omega 3s and 6s, are available in capsule form from health-food stores and pharmacies.**

KEY FACTS

* When its cell walls lack fats, the skin's ability to protect itself against sun damage is reduced.
* Regularly eat cold-pressed, unrefined vegetable oils and oily fish, or take a course of evening-primrose oil or borage oil in capsule form.

07 don't forget vitamin E

Vitamin E protects against the sun's rays through its strong antioxidant powers. It also increases the capacity of cells to repair themselves and improves the general condition of the skin. Unfortunately, a healthy, balanced diet doesn't always provide us with as much as we need.

Precious vitamin E

When free radicals increase as a result of the body's exposure to sunshine, the membranes and the core of skin cells are both damaged. Vitamin E limits this damage, so we must ensure our skin gets enough of it all year round.
Nuts, seeds and vegetable oils are good sources of this vitamin, but they only contain small quantities and it's not

●●● DID YOU KNOW?

> Vitamin E is even more effective as an antioxidant when combined with vitamin C.
> Vitamin C is found in numerous plants, including kiwis, citrus fruits, soft fruits such as strawberries and raspberries, leafy green vegetables like spinach and water-cress, and cabbages.
> As vitamin C stimulates the immune system, it helps to prevent cancers caused by overexposure to the sun.

always possible for us to extract an adequate amount from them, because the plants themselves use it to reduce the oxidation of their fats.

Wheatgerm oil

An unstressed person living in an unpolluted environment who doesn't smoke and doesn't take the contraceptive pill can rely on a balanced diet to provide all the vitamin E that is required. A teaspoonful of wheatgerm oil, for instance, containing 1 mg of vitamin E, would be adequate. But, these days, who can claim to be totally stress- and pollution-free? You might need as much as 10 or even 15 mg per day, so taking supplements (capsules or tablets) may be the best course of action. However, it should be noted that only vitamin E from natural sources can be metabolized by the body.

> You can take vitamin C in tablet form, but it is more effective when taken from natural sources (such as dog-rose or Barbados cherry), as the body absorbs it more efficiently.

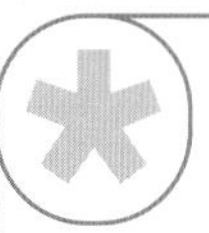

KEY FACTS

* Vitamin E is an essential defence against the damage that sunshine can cause to skin cells.
* Vitamin E works better when combined with vitamin C.
* A healthy, balanced diet doesn't always provide us with all the Vitamin E we need.

08

The tomato is a sun plant par excellence. Not only does it grow in hot and dry conditions, but it can also protect the skin from sun damage. One of its key constituents is lycopene, which is a powerful antioxidant.

tomatoes have a lot to offer

The benefits of lycopene

The tomato is the quintessential summer fruit, although it is eaten as a salad rather than a dessert. It also provides our skin with wonderful protection from the sun's rays.

When applied directly to the skin, tomato pulp cools and soothes sunburn, but, most importantly, it contains large quantities of lycopene. This is a vigorous

DID YOU KNOW?

> Unlike most vitamins, lycopene is not damaged by being cooked. In fact, cooking makes it more easily absorbed.

> When you eat raw tomatoes, you absorb only 10 per cent of the lycopene; but when you cook them, 70 per cent of it is assimilated by the body.

> If you add unrefined oil (rape oil, olive oil, etc.) to tomatoes, the percentage that is absorbed rises even further.

> You can also grind or liquidize them or put them in a juice extractor to liberate the lycopene.

antioxidant that is found only in the tomato (and, in small quantities, in watermelons and pink grapefruits). So you should eat tomatoes frequently to reap its benefits. And what benefits they are. Several studies have shown that it protects all the tissues from free radicals: the joints, the muscles, the brain and, of course, the skin.

A natural product

Lycopene, like beta-carotene, belongs to the carotenoid family. It combines with the antioxidant vitamins (A, C and E) to neutralize, in particular, the destructive free radical singlet oxygen. As the tomato also contains vitamin C (17 mg per 100 g/4 oz), vitamin E (1 mg per 100 g) and beta-carotene (600 mg per 100 g), it's fantastically beneficial.

Eat some every day, because there are no downsides: tomatoes contain few calories, are a natural diuretic and protect against excess cholesterol. What's more, there are several varieties on the market, each with a subtly different flavour, so there's no need to get bored.

KEY FACTS

* When applied directly to the skin, tomato pulp cools and soothes sunburn.

* Tomatoes contain lycopene, a major antioxidant, as well as important vitamins E and C and beta-carotene.

* They help provide natural protection against sun damage.

Maize, cabbages and asparagus contain important sun-protection substances. These three rare but vital antioxidants are: zeaxanthin, which protects the eyes; lutein, which prevents sun allergies; and glutathione, which improves the skin's immune defences.

09 you can rely on this trio!

Marvellous broccoli

Cruciferous vegetables are very rich in lutein. Numerous studies have shown that one of them – broccoli – acts especially well against certain cancers because of its powerful antioxidant qualities. When it comes to sunbathing, lutein quickly repairs DNA damaged by the sun's rays. All crucifers, even those that contain only small amounts of lutein

●●● DID YOU KNOW?

> Zeaxanthin is an antioxidant that improves the general condition of the retina and its ability to recover from damage.

> This is important because exposure to the sun affects the eyes: they can suffer significant damage even when you keep your eyelids closed.

> Maize (sweetcorn) is the best source of zeaxanthin. In autumn, eat it raw, boiled or grilled with a little butter. (Butter contains vitamin A, which complements the action of the zeaxanthin.) The rest of the year, eat canned sweetcorn.

(watercress, radishes and turnips, for instance) are recommended, particularly for people who suffer from summer polymorphic light eruption (see Tips 58 and 59).

Indispensable asparagus

Asparagus contains a sulphur amino acid called glutathione, which prevents the sun from disrupting the skin's immune reactions. It also protects the eyes against cataracts. When it's in season, eat plenty of this delicious vegetable in order to build up your reserves. During the rest of the year, garlic, onions and egg yolk will do instead.
Note: both asparagus and broccoli also contain a lot of uric acid.

> Don't be concerned about consuming too many calories. Maize is certainly high in calories (122 per 100 g/4 oz tin), but it fills you up, so you'll be less likely to snack.

KEY FACTS

* Cruciferous vegetables contain lutein, which helps repair DNA damaged by the sun's rays.

* Asparagus is rich in glutathione, which prevents the sun disrupting the skin's immune responses.

* Sweetcorn is a source zeaxanthin, which protects the eyes.

10 selenium and zinc, a high-power duo

You can't fully protect yourself from sun damage without eating an adequate amount of selenium and zinc. A healthy, well-balanced diet should provide you with all you need, so eat plenty of fish, shellfish, eggs ... and Brazil nuts.

Toxic or therapeutic?

For more than a hundred years, selenium was considered to be toxic. It was thought that eating too much of it could cause digestive or skin problems, for example. It was only about twenty years ago that its true biological significance and benefits were discovered. It is vital to the human body, although only in tiny quantities: we need something like 60 to

DID YOU KNOW?

> The Brazil nut contains more selenium than any other fruit or vegetable. Other good sources are shellfish, meat and eggs.

> Selenium works in combination with vitamins A, C and E by improving their antioxidant properties.

> Zinc is found in liver, almonds, fish and shellfish.

> The phytates in certain plants inhibit the body's ability to absorb zinc, so it's better to rely on animal products for a supply of this mineral.

80 mcg per day. As the food we eat contains very little selenium, however, there's no risk of overdosing, even if we eat food 'rich' in it. Its main function is to prevent the oxidation of fatty acids by helping glutathione do its work (see Tip 9), and we know how vital fatty acids are for the health of our cells in general and our skin cells in particular. If your selenium level is too low, you are at greater risk of suffering skin damage from the sun's rays.

You can rely on zinc

Zinc is involved in many biological reactions. As far as the skin is concerned, it's needed to soothe inflammation, accelerate repairs to damaged areas, maintain immunity and ensure the successful working of hormones. A shortage can cause, among other things, dermatitis, slow-healing wounds and loss of local pigmentation.
Don't sunbathe if you are suffering from a lack of zinc: it would just be asking for trouble. We need about 15 to 20 mg per day, which could be provided, for example, by 100 g (4 oz) of oysters.

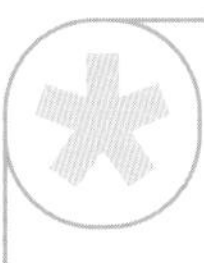

KEY FACTS

* Selenium helps prevent the oxidation of fatty acids in the membranes of skin cells.

* Zinc enhances all the skin's biological functions.

* Including fish and shellfish in your diet should provide you with enough of these essential minerals.

11

try 'pre-sun' tablets

Pre-sun tablets contain antioxidants that can help the body's natural protective system to kill the free radicals produced when ultraviolet (UV) light strikes the skin. They can be effective, provided you take them early enough.

The basic ingredients

In spring, a look at the shelves in a pharmacy will reveal a wide variety of suncare products. Among the tanning creams and after-sun lotions, you can also find pre-sun tablets designed to prepare the skin for the summer by providing it with all the necessary nutrients. If you compare what's in the different products, you'll see that most contain the same basic ingredients: beta-

DID YOU KNOW?

> Taking supplements doesn't mean you can overlook all the usual precautions you should take when sunbathing.

> They can't take the place of protective creams, nor eliminate the need to acclimatize gradually to the sun.

> They are unsuitable for children, who shouldn't sunbathe in any case.

> Food supplements containing high doses of beta-carotene sometimes give an orange tint to a suntan, particularly if you take a lot of them.

carotene, essential fatty acids and, often, vitamins A, C and E. Each laboratory then adds its own special touch: lycopene, tea extract, various minerals and so on. These supplements work well, provided you take them two weeks, or preferably a month, before you start sunbathing. You need to take them for a cetain amount of time in order to build up your reserves.

Keep topped up

These products should complement a well-balanced diet. Many people's diets are seriously lacking in vitamins, particularly E and C, as well as selenium and zinc (see Tip 10), and taking some capsules for a few days won't be enough to make good these deficits. You should continue taking the supplements throughout your holiday period, or the reserves built up before you went away will soon be exhausted. Finally, don't forget to continue taking the tablets for two to four weeks after you return home, to help your body repair any sun damage that may have occurred.

KEY FACTS

* Pre-sun tablets contain beta-carotene, vitamins A, C and E, fatty acids and in some cases plant extracts.

* They only work if you take them for at least two weeks before starting sunbathing and two weeks after, but don't neglect your other sun precautions.

12 take a course of horsetail

Horsetail contains a mineral that helps to keep the body's connective tissues in good condition. Your bronzed skin will thank you for taking it.

Proteins, gel and fibres Our skin is partly made up of connective tissue. This consists of cells held together by collagen, a substance composed of proteins, gel and fibres. Collagen helps to ensure that the skin remains elastic, supple and moist. The sun harms connective tissue, so it needs to be strengthened and aided when recovering from any damage.

A rich source of silicium Horsetail is full of silicates, indispensable for regenerating connective tissue, as well as vitamin C and glutathione, which help strengthen our antioxidant defences. The silicates in horsetail are easily absorbed by the body, which is not true of silicium with a mineral origin. It has other benefits, too, including being good for your bones. However, a few words of caution: horsetail should not be taken in pregnancy or for extended periods of time and you must always obtain it from a reputable supplier, as it must be specially treated to deactivate an enzyme that interferes with vitamin B1 (thiamine) in the body.

●●● DID YOU KNOW?

> Horsetail is available in capsule form: take a three-week course before going away on holiday or starting any sunbathing.

> Try an infusion: use 30 g (1 oz) of the dried plant per litre (2 pints) of cold water; boil for 15 minutes, then leave to infuse for 10 minutes. Drink three cups a day for three weeks.

KEY FACTS

* The skin's condition relys on the state of its connective tissue.

* Exposure to the sun damages the connective tissue, so take a course of horsetail three weeks before you start sunbathing.

13 diet with caution

In the month before going on holiday, we often diet to lose those few extra kilos before donning a swimming costume. This doesn't help the skin, though.

Slimming yes; deficiencies no Almost everyone tries to shed a few kilos before heading for the beach. However, be careful not to deprive your body of any of the substances it will need. If you have been eating a sensible, well-balanced diet throughout the rest of the year, your body will be able to withstand a few deprivations without coming to any harm. But if you haven't, don't start a rigorous diet a month before going on holiday.

Diet sensibly A sudden, excessive reduction of food intake will lower your reserves of vitamins and minerals and could harm your skin. If you then start sunbathing, your body will not be able to offer you much protection. Instead, cut back gently on the amount you eat, starting three months before your trip. Ensure you don't deprive yourself of foods essential to the skin, such as fruits, vegetables, cereals, fish and unrefined oils.

DID YOU KNOW?

> You should plan to start slimming three months before you go away on holiday, and combine your regime with a course of vitamins and minerals.

> First, try standard multi-nutrient supplements; then switch to a special pre-sun course a month before you leave.

KEY FACTS

* Avoid severe dieting before going away on holiday.

* If your skin is deprived of essential nutrients, it will be unable to protect itself from sun damage.

14

a nice cup of tea

Tea is brimful of antioxidants, which protect the skin from sun damage. Whether you drink it cold or hot, whether you prefer green or black, it's an ideal summer drink. It's also very rich in zinc and vitamins. There are plenty of flavours to enjoy, too: flowers, fruits, spices ...

The most popular drink in the world

The leaves of the plant *Camellia sinensis* are used in infusions all over the world. More tea is drunk than any other beverage. It is enjoyed everywhere, from English tearooms to the burning sands of the Sahara, the misty mountains of Tibet and the Russian steppes. It tastes good, but another side of its appeal is that it is also a medicinal plant, with a great deal to offer the serious sunbather. For

●●● DID YOU KNOW?

> Black and green tea both come from the same plant. The colour and taste variations are due to different production methods.

> Black tea is rolled to break the veins in the leaves and then fermented before being dried. Green tea is dried as soon as it is picked. Their properties are slightly different, but both teas are antioxidants and protect the skin.

> White tea, which comprises very young leaf buds picked before they are ripe, is only produced in small quantities and is therefore much more expensive than other varieties.

instance, it contains a large quantity of flavonoids, substances with strong antioxidant properties of their own that also increase the action of other antioxidants, particularly vitamin C.

Inside and out

Tea also contains vitamins E and A, as well as zinc, all substances that have a direct effect on the skin. If you drink tea regularly, it will help keep your skin supple and soft, even after exposure to the sun, and, even more importantly, will protect you from long-term damage to your skin cells. If tea causes you to feel nervous and edgy, let it brew for a long time: the plant's tannins, which spread slowly into the hot water and give the drink its dark colour, neutralize the caffeine, which is responsible for its stimulant effect.

If you have some tea left in your pot, don't throw it away: it makes an excellent lotion that will close your pores, soothe sunburn and, if it's black tea, even slightly enhance your tan.

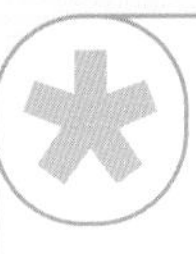

KEY FACTS

* Whether green, black or white, tea is a wonderful health drink.

* It is very rich in flavonoids, zinc and vitamins E and A, as well as antioxidants that protect the skin from the sun.

* You can also use cold tea, after it has been infused, as a soothing skin lotion.

15

sunbathing and skin problems

Now you've read the first 14 Tips, do you think you're ready for some serious sunbathing? Not necessarily. People with certain skin types or specific skin conditions, such as rosacea and vitiligo, need to be very careful and should sometimes avoid exposure to the sun all together. Moles can be a problem, too.

The white …

If you have a very fair complexion, fair hair and blue eyes, and the slightest exposure to sunshine makes you look like a lobster, it's probably best simply not to bother. Even if you prepare and protect your skin carefully, you will never end up with much of a tan. If you have to go out in the sunshine, put on a very protective cream or even a total

DID YOU KNOW?

> Closely examine your beauty spots and moles regularly for any changes. The lighter your complexion, the more necessary it is to keep a close eye on them.

> It's no good only checking when you return from holiday, as malignant melanoma (a form of cancer) does not arise only after a heavy bout of sunbathing, as many people think. Melanomas develop as a result of the cumulative effect of exposure to the sun over many years.

> If you have a beauty spot or mole that changes shape, size or texture, consult a doctor immediately.

sun block. Avoid sunbathing and, when you're on the beach, stay under a sunshade or wear a T-shirt.

You should also avoid sunbathing if you suffer from vitiligo – white patches, usually on the face, hands or neck, due to a loss of pigmentation. This occurs when the melanocytes die, meaning the skin is unable to produce melanin. If you expose those white patches to the sun, they will not disappear, but worse, they will actually stand out more prominently from the rest of the skin. Since the skin on the patches is unprotected, it will burn very quickly. In summer, always ensure you put total sun block on the affected areas.

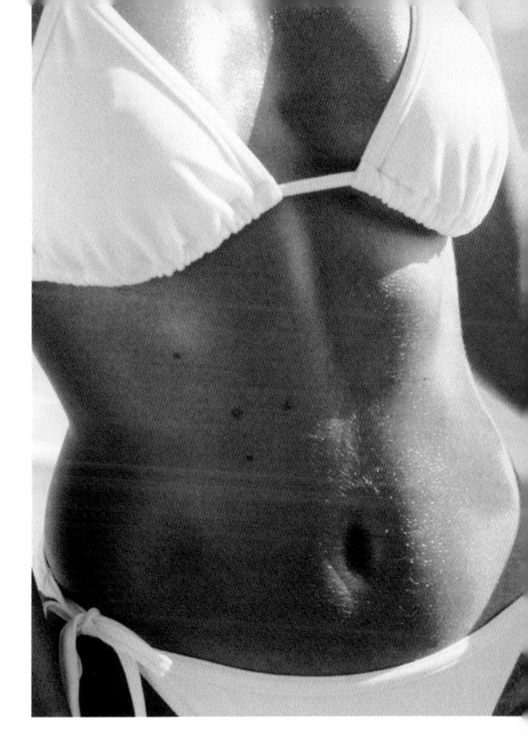

… and the pink

Rosacea also tends to worsen in the sun. These networks of fine blood vessels, which generally appear on the face, are caused by a circulatory problem. Exposure to the sun never improves blood circulation, so put a very high-factor sun cream on your face before going out and always wear a wide-brimmed hat.

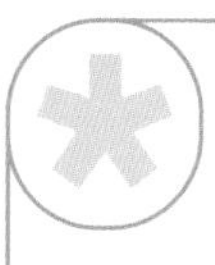

KEY FACTS

* If you have a very fair complexion, you should avoid exposing it to the sun as much as possible.

* White patches caused by vitiligo are particularly vulnerable to the sun's rays.

* Don't expose rosacea or beauty spots or moles to bright sunlight for long periods.

16 avoid the sun if you have acne

Every teenager will probably tell you that sunshine makes acne disappear. However, no matter how much they argue that lying around in the sun all day is good for them, that's not the case. In fact, it's best to avoid long periods in the sun and to use a high-protection sun cream if you suffer from acne.

The pros and cons of teenage sunbathing

Sunlight helps reduce the appearance of acne because it suppresses some of the inflammatory reactions in the skin. This means that the spots appear less red and prominent. But sunlight can sometimes worsen acne by causing scaling that blocks the follicles in the skin.

●●● DID YOU KNOW?

> Burdock, with its purifying properties, is very effective against acne.

> Follow a course of treatment with a base of burdock during the three weeks before your holiday.

> For internal use: add 60 g (2 oz) of dried burdock to 1 litre (2 pints) of cold water, bring to the boil and allow to boil for 15 minutes; leave to infuse for another 15 minutes. Drink three cups per day.

Zinc and fruit acids

Protect the areas affected by acne with non-greasy, high-factor sun cream and wash your skin even more carefully than usual after every sunbathing session. Common medical treatments for acne often contain photosensitive substances (see Tip 19). For example, sunlight may trigger a rash in patients using tetracycline antibiotics. Check carefully to see whether your acne treatment is safe to use in sunlight or consult your pharmacist. You can continue to use products with a base of fruit acids or zinc without risk, but avoid the others.

> **For external use: add 15 g (1/2 oz) of burdock root and 15 g (1/2 oz) of wild pansy to 250 ml (1/2 pint) of water and bring to the boil. As soon as the mixture boils, take it off the heat and leave to infuse for 10 minutes. Clean your spots with this lotion morning and evening.**

KEY FACTS

* The sun may cause acne spots to appear to diminish, but it is ceratinly not a 'cure'.

* Use high-factor protection cream on the affected areas and medication with a burdock base.

* Common treatments for acne often contain photosensitive substances.

17

beware of sun blisters

If your lips are affected by unsightly blisters as soon as you go out in the sun, it's due to a virus activated by the sun's rays. These blisters can be dealt with in two stages – by taking certain plant extracts before sunbathing to strengthen your immunity, and by protecting the areas at risk with total sun block.

Sun and immunity

Once the herpes virus enters your body, it never leaves: it stays hidden inside you. However, it may never be activated again: it all depends on how effectively your immune system deals with this enemy within. You can take comfort from the fact that, although a large majority of the population carries this virus, it can be neutralized so that it remains dormant or is active only seldom.

●●● DID YOU KNOW?

> The sun's rays threaten not only the lips but any area of exposed skin. Try an immunity-boosting treatment before going on holiday.

> Homeopathic treatments are a good long-term way of suppressing herpes. To obtain them, you need to consult a homeopathic practitioner.

> You may find an acyclovir cream, available from pharmacies, helpful in combatting these painful sores.

> Alternatively, an antiviral antibiotic, taken on a long-term basis to suppress the herpes virus, could be prescribed by your doctor.

Exposure to the sun can stimulate the virus into life: the production of singlet oxygen free radicals by solar rays damages the body's tissues and rapidly reduces immunity. Once the immune system drops its guard, the hitherto dormant virus attacks.

Echinacea and sun block

To avoid these unsightly and painful sores, don't expose yourself to the sun too suddenly (prepare yourself over several weeks), don't stay in the sun for too long and always apply a high-factor protective sun cream and a specialist sun-stick for your lips.

Echinacea strengthens the body's immune defences. It is available in capsule form or as a liquid extract. Remember to start the treatment some time before going away.

The shiitake mushroom is another effective booster to the immune-system. Consuming them will put your body in a better position to withstand sun damage. During the first few days of your holiday, increase the time you spend in the sun very gradually and apply total sun block to the areas at risk, such as around the eyes and mouth and on the lips. This will prevent the harmful rays from penetrating and help keep your immune levels intact.

KEY FACTS

* Sudden, prolonged exposure to the sun can reduce your immunity level and activate the herpes virus.

* To avoid this, try to boost your immunity for several weeks before starting to sunbathe.

* Use a total sun block on areas that are at risk.

18 discourage those brown blotches

Some women find that brown blotches appear on the skin as soon as it is exposed to the sun. Mainly affecting pregnant women and those taking the contraceptive pill, these blotches are caused by oestrogen acting on the skin. Protecting your skin is the only way to avoid them.

DID YOU KNOW?

> If you've done a lot of sunbathing over the years, you must start protecting yourself immediately.

> When you go to the beach, wear a hat to shelter your face from the direct rays of the sun, stay in the shade in the middle of the day and use a total sun block.

Chloasma

Pregnant women are familiar with these brown blotches, which appear on the face (particularly around the mouth and eyes) and on the stomach and breasts (around the nipple). They are due to an increase in pigmentation. The body's hormones, which undergo substantial changes during pregnancy, cause melanocytes to become hyperactive. Melanin then accumulates to form blotches. They usually disappear spontaneously three to six months after delivery, unless you have done a lot of sunbathing during your pregnancy. The effect of the UV rays combined with the hormones can make these clusters of melanin permanent. Women taking contraceptive pills containing oestrogen can be similarly affected.

Age spots

The same sort of blotches appear on many women around the age of fifty. Known as 'age spots' even though they appear long before old age, they particularly affect people who have spent many years sunbathing or walking about in the sun with their skin unprotected. They occur on areas frequently exposed to the sun, such as the face, arms and shoulders. Areas that are usually kept covered, such as the breasts and buttocks, are generally spared. Those who have been sensible throughout their lives by wearing broad-brimmed hats and religiously applying sun cream will tend to reap the benefits once they reach fifty. Those who took no such precautions will almost inevitably be the first to be struck by 'age spots'.

> There's only one way to prevent chloasma: stay out of the sun as much as possible and apply sun block when do you have to venture outside.

KEY FACTS

* Chloasma (brown blotches) can become permanent.

* Avoid long periods of sunbathing and protect yourself well.

* Age spots can appear if you've sunbathed regularly for years.

* Don't forget to use a total sun block.

19

steer clear of bergamot

Some substances are photosensitive: they increase the skin's sensitivity to the sun's rays, causing spots and blotches and can sometimes result in eczema and oedemas. So beware of bergamot and certain perfumes and drugs.

Like a magnifying glass

The sun's capacity to burn the skin depends on what type of skin we have and also on the products we put on it. Some creams can protect the skin (see Tips 26 and 27) but other products have the opposite effect, accelerating the rate at which the skin burns, to the extent of causing red patches, itchy rashes, or even eczema.

DID YOU KNOW?

> If you have sensitive skin or suffer from skin allergies, always make sure you choose products specially made to avoid allergic reactions and which are free of photosensitive substances.

> Even tanning creams sometimes contain colourings that can cause a reaction in very fragile skin. Read the labels carefully and ask your pharmacist for advice.

> Sun allergies (see Tips 58 and 59) can be exacerbated by photosensitive products.

Oedemas (swellings due to the build-up of fluid) can also occur in the area where the products have been applied.
If there is a major adverse reaction, the effects could extend to the whole body. Sometimes, fortunately, the result will only be brown blotches, which, although unsightly, are not serious.

Drugs and cosmetics

The most well known of these substances is bergamot. Extract of this plant produces brown marks that are very difficult to ghet rid of. But bergamot does not produce the worst effects. Some drugs, such as anti-inflammatories, antibiotics, neuroleptics, antidepressants and tranquillizers, can have severe side-effects. If you are taking a course of drugs, check on the label to see whether they're likely to cause a problem and/or ask your doctor or pharmacist for advice.
Some cosmetics can also cause reactions, particularly perfumes or their by-products, such as perfumed lotions, creams and talcum powders.

KEY FACTS

* Some substances, such as bergamot, and certain drugs and cosmetics, cause brown patches and even rashes, eczema and oedemas.

* If you have especially sensitive skin, choose products specially manufactured to avoid allergic reactions.

20 protect your scars

A surgical operation or an accident could leave you with scars. If you want them to fade, don't expose them to direct sunlight. Extensive scar tissue, such as that caused by a burn, lacks melanocytes and will not tan.

Recent scars When the skin has been cut, because of an accident or an operation, it needs time to mend. Exposure to the sun, which disturbs the skin's functions in general, interferes with the healing process. Scars will become more prominent and are likely to remain so after your sunbathing session is over. If they tan at all, they may later fade to a mottled appearance.

Six months, a year, two years ... If your scars are recent (less than six months old), under no circumstances expose them to the sun. Keep them covered or keep in the shade. If you've had them for six months to a year, protect them with a total sun block to keep contact with the sun's rays to a minimum. It is possible to expose scars to the sun after two years, but always make sure you protect them with non-photosensitive products.

DID YOU KNOW?

> If you are able to decide the timing of an operation, try to avoid the spring.

> Choose autumn or the beginning of winter instead, to avoid all possibility of the scars being exposed to the sun.

> Take care if you undergo 'vigorous' cosmetic treatment, such as dermabrasions or facial scrubs.

KEY FACTS

* Recent scars should never be exposed to the sun.

* Scars can be exposed after a period of two years, but protect them with a non-photosensitive cream.

case study

No more dieting for me before going on holiday

«I always used to diet before going on holiday, to lose the 2 or 3 kilos (4 to 6 lbs) I'd put on during the winter. After my pregnancy, I decided to lose ten or so kilos (22 lbs) the same way. I kept to a very severe diet: I ate very little and completely cut out fruit, fats and cereals. I stuck to salads without any dressing and grilled meat. I kept this up for three months. I was happy when I went on holiday, because I'd achieved what I'd set out to do. The following week, however, I realized that something was badly wrong: I had sunburn like never before and then itchy rashes and eczema. What's more, herpes, which hadn't bothered me since I was about fifteen, suddenly flared up again. I just couldn't understand why it was happening, but a doctor, a specialist in nutrition, explained it all to me later: I was short of vital nutrients like vitamins, minerals and fatty acids. These days, I prepare for my holidays by taking care to eat a very different kind of diet.»

21 »»

» **The sun must be treated with a great deal of respect.** Be sensible, don't rush it and always take the right precautions. Only then will you be able to get a suntan safely.

»» **Modern sun creams provide very effective protection** against sunburn, but even they won't help you if you sunbathe for too long during the first few days of your holiday.

»»» So choose your creams well, be sensible about how much time you spend in the sun, protect the sensitive parts of your body (especially your eyes) … **and always take twice as much care when dealing with children**.

40
TIPS

21 enjoy the benefits of the sun

Let's make one thing immediately clear: the sun is not our enemy. Without it, there would be no life on earth. It gives us light and heat. It orchestrates our biological life, regulates our moods and even strengthens our bones. And it's good for our skin, as long as we treat it with the respect it deserves.

The great conductor of the rhythms of life

With the sun, it's all a question of amount: too much of a good thing is bad for you. All our vital functions are conditioned by the rhythms of night and day and the changing of the seasons. Lack of sunlight has a series of adverse effects: a reduction in libido and alertness, tiredness, sleepiness and depression. The sun's rays travel from the eyes, via the retina and the optic nerve, to a part of

DID YOU KNOW?

> The synthesis of vitamin D, which helps strengthen bones, occurs thanks to the sun. The skin produces this precious vitamin in response to the sun's rays, although you can also obtain it through your diet.

> If you don't go out often in the sun, you could suffer from a lack of vitamin D.

> It is rare for adults in the Western world to have a vitamin D deficiency, but it can affect babies, young children and elderly people, as well as Muslim women who wear traditional costume.

the brain called the hypothalamus, which controls the pituitary gland. This gland is known as the 'leader of the endocrine orchestra' as it produces hormones that stimulate the thyroid and adrenal glands, as well as the ovary and testis, and it therefore affects our whole body.

Life in the open air

In the nineteenth century, 'sun therapy' was a common treatment for a variety of ailments. In the following century, such therapy was widely dismissed as quackery, but now we are beginning to appreciate its value again. Health centres often include solariums, where you can lie in the sun at specific hours of the day for a limited amount of time. The sun certainly helps to keep the skin in good condition, provided, of course, you don't have too much of it. But what is the right amount? The answer is roughly the amount you would get from living in the open air in a moderately sunny country without sunbathing. This is less than the amount needed to get a 'good' tan.

> The solution is to go for walks when the weather is sunny or take supplements. A 15-minute walk around midday in short sleeves on a sunny, spring day is enough to provide a good synthesis of vitamin D for 24 hours.

KEY FACTS

* The sun is essentail to all forms of life, as it provides light and heat.
* Its rays enter the body through the eyes and regulate all our internal functions.
* Lack of sunshine causes tiredness, a decrease in libido and alertness, and depression.

22 when the weather gets warmer...

To grow accustomed to sunshine gradually, do a little sunbathing on the balcony or in the garden as soon as the weather becomes warmer. Your skin will slowly adjust to the sun and your body will gradually start to produce melanin again.

After hibernation

A tan that is developed gradually will last a lot longer. Your skin needs time to establish its defences, and the more time you allow it, the more it will repay you with a golden complexion, while still remaining healthy, supple and soft. Your melanocytes have been sleeping all winter. In high summer, if you suddenly bombard them with rays, they won't be able to cope with all the demands being

DID YOU KNOW?

> If you live in a temperate climate with clearly defined seasons of transition (spring and autumn), you should take advantage of the fact.

> Benefit from the sun (30-minute sunbathing periods once or twice a week) when the temperature is around 20°C (68°F).

> From the beginning of June in the northern hemisphere, however, start to take more care, because the sun starts to become more intense.

> If you have sensitive skin, protect yourself with a medium-factor sun cream even in the springtime.

made on them and will allow destructive UV rays to get through. So start them off gently in the spring sunshine. If you sunbathe for a short time occasionally but regularly (once or twice a week), they'll be better able to deal with the intense sun of the holiday months.

Before the summer sun

Even if it's still not very hot, get out and about at the first sign of sunshine. Don't take the car or the underground but instead walk on the sunny side of the street. At the weekends, take a stroll in the country rather than going to the cinema. Get a little sun on your arms and legs.

If you have a garden or balcony, put on your bathing costume and lie out in the spring sunshine for 30 minutes (but no more). During spring, the sun is further from the Earth than in summer and is therefore less intense: its UV rays are less powerful, so you will gently ease the melanocytes into life. Even if your skin shows little sign of a tan, these cells will start to stir and begin their protective duties. Your skin will slowly become acclimatized and so will be much less vulnerable when it is exposed to the summer sun.

KEY FACTS

* For gradual, gentle tanning, nothing beats the spring sunshine: in the garden, on the balcony, during country walks.

* But don't rush things: no more than 30 minutes once or twice a week, and use a medium-factor sun cream if you have a sensitive or delicate skin.

23

avoid the hottest part of the day

The most important rule of all is don't sunbathe during the hottest part of the day in summer. This is not only to avoid sunburn, but also because, even if you've already got a deep tan, the sun's powerful rays around the middle of the day can still cause invisible damage.

The most common mistake

Most people are wise enough to sunbathe relatively carefully during the first few days of their holiday in order to avoid the pain, headaches and high temperatures caused by sunburn. However, as soon as their skin browns and protects itself from the UV rays that cause sunburn, many sunbathers become careless. The midday sun in high summer is

DID YOU KNOW?

> If you start to get a tan after just a day or two, be warned: it's a 'false tan'.

> The UV rays cause the melanin already present in your skin to oxidize and rapidly give you some colour, but this provides no extra protection.

> You'll have to wait for about a week before new melanin is available.

> This provides a deeper colour and is longer lasting. It lasts for about two months, in comparison to the 'false tan', which fades after a few hours.

still harmful even after you've got a deep tan. Some of the UV rays continue to penetrate right into the cells, where they stimulate the production of destructive free radicals, which accelerate the ageing of the body's tissues.

Never in the midday sun

The best solution is the simplest: never sunbathe during the hottest time of the day. The sun should be avoided completely between eleven and three in many countries, even in the UK. As for children and people with sensitive skin, or those going to hot destinations such as Greece and Spain in high summer, stay in the shade at all times after mid-morning and before late afternoon.

A good rule of thumb is to stand in the sun for a moment and look at your shadow. If the shadow is shorter than your actual height, the sun is too high in the sky for safe sunbathing. If you are spending the whole day outside (on the beach, in the country or in the mountains), stay in the shade as much as possible during this part of the day. (A tree provides better protection than a sunshade.) Wear a broad-brimmed hat and a shirt.

KEY FACTS

* Never sunbathe during the hottest hours of the day.

* The powerful rays that bombard you at this time of the day are harmful, even if you have a tan.

24 screen those dangerous UV rays

Sunlight is made up of a spectrum of rays, some of which are invisible. The most harmful ones are blocked by the earth's upper atmosphere, but those that reach the surface all have their own special characteristics. UVA and UVB rays are the most significant when you are sunbathing.

Gamma, infrared and X rays

The sun bombards us with many different rays, the most dangerous of which are X and gamma rays, but they are prevented from reaching us by the ozone layer, which is about 30 kilometres (20 miles) above sea level. If we continue as we are now, and destroy the ozone layer completely, they might get through unhindered, which would be disastrous. But we are not at that stage quite yet.

DID YOU KNOW?

> Since UVA rays penetrate to the very core of the body's cells, they can damage the DNA molecules that carry genetic information.

> Once damaged, these molecules can no longer reproduce normally and, if not eliminated from the body, can increase the risk of cancer.

> People with red hair and/or lots of freckles are the most vulnerable, because the red melanin they produce in large quantities seems to offer less protection against ultraviolet rays and may even increase the risk of cell mutation.

Infrared rays provide the Earth's heat. They do no harm to the skin, but too many of them can cause heatstroke and sunstroke (see Tips 50 and 51).

Dangerous ultraviolets

Ultraviolet rays are divided into three categories: UVAs, UVBs and UVCs.
UVCs, the most dangerous of all, are absorbed by the ozone layer.
UVBs cause sunburn, but melanin gives us some protection from them. When we get a deep tan, we reduce the risk of being burned, but the protection given is slight – about the same as a low-factor sun cream. Only 20 to 30 per cent of the sun's UVB rays reach Earth, but that's enough to cause burns, skin ageing and wrinkles. They are also implicated in some non-melanoma skin cancers.
UVA rays can also produce sunburn, but to a lesser extent than UVBs. They cause immediate tanning by activating melanin in the skin. This can cause serious problems as the more tanned we are, the fewer precautions we tend to take. UVAs penetrate deep into the skin, even if it is well bronzed, and contribute to its premature ageing. Although there's currently no consensus on which type of UV causes melanoma, many suspect it to be UVA. About 80 per cent of the UVA rays emitted by the sun reach the Earth's surface.

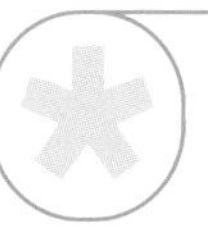

KEY FACTS

* Ultraviolet rays are extremely harmful.

* UVB rays cause sunburn but are blocked by melanin. A deep tan protects us from them, but only like a low-factor sun cream.

* Even if you have a suntan, UVA rays can still damage the cells.

25 a few facts about sun creams

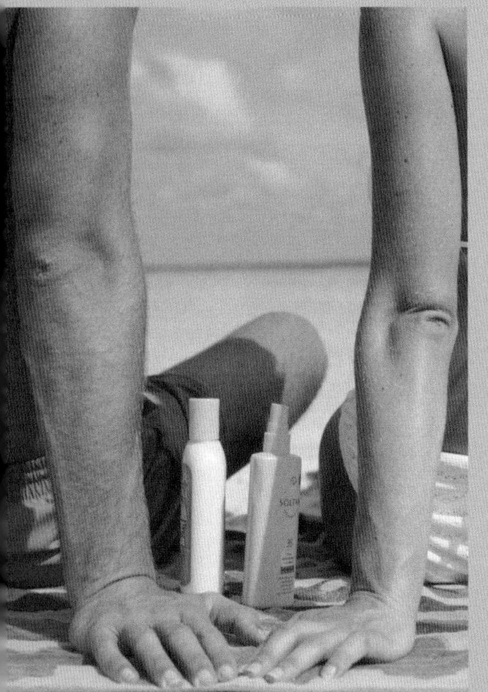

Protective creams have come a long way in the past twenty years. Nowadays, they contain effective chemical or mineral filters and there is a wide range of carefully graduated protection factors available.

DID YOU KNOW?

> There are also ranges of organic sun-care products available.

> These products do not contain any photosensitive substances or chemical filters and all their ingredients are natural.

> They display protection factors that conform to current legislation.

Choosing the right factor

The labels on sun-care products sometimes seem impossibly complicated. There are so many figures and factors that it can get confusing. So here are some pointers to help you understand these products better.

First, the protection factors. They range from 2 to 100: the higher the number, the greater the protection. But it's not as simple as factor 50 giving twice as much protection as factor 25. A factor 2 cream blocks 50 per cent of UV rays; factor 20 blocks 95 per cent; factor 50 blocks 97 per cent; and factor 75 blocks 98 per cent. These factors usually refer to protection from UVB rays. The factor numbers are calculated as follows: if you have applied a factor 10 sun cream, after ten hours in the sun you will have absorbed the same amount of UVB as you would have done in one hour without protection. Apply factor 50 and theoretically you can stay in the sun for fifty hours.

There's also a star system to indicate UVA protection. One star is the lowest protection factor and four the highest.

> It is recommended that you always choose a cream that will protect you from both UVA and UVB rays.

Chemicals or minerals?

Manufacturers use two types of filter in their sun creams. Chemical filters contain synthetic substances that penetrate the skin. They are effective but can cause adverse reactions, so avoid them if you have sensitive skin or suffer from any allergies.

Mineral filters do not penetrate the skin. Instead, they deposit a thin layer of protective minerals, such as zinc, talcum and mica, on its surface. They do not cause allergic reactions and, unlike chemical filters, whose active ingredients are gradually neutralized by the sunlight, they remain effective for a long time.

KEY FACTS

* The higher a sun cream's protection factor, the greater the protection.
* Chemical filters penetrate the skin, whereas mineral filters remain on the surface.
* Organic sun-care products are also available.

26 choose the right factor

Before going on holiday, always pack several sun creams with different protection factors. As the days go by, you will be able to gravitate to less protective creams. But don't rush things, and always bear in mind your skin type before choosing which cream to use.

Which type?

Should you choose a cream, an oil, a lotion, a stick or an aerosol? If you have a dark complexion, choose a lotion or an oil. Oils offer less protection but are more water resistant. Like lotions, they can be bought in aerosol form, which are easier to use but more expensive. You can use creams on your face and sticks on sensitive areas, such as the mouth,

DID YOU KNOW?

> The amount of protection you need also depends upon where you are.

> Tropical regions are particularly hard on the skin, as the solar rays there are two to ten times more powerful than in England.

> When your skin has had enough sunshine for the day, go in the shade (preferably under a tree) or put on a sarong or T-shirt. But don't think this is completely blocking out the sun: its rays can pass right through most fabrics.

> Remember that sand and snow reflect UV rays very well, even if you are in the shade.

around the eyes, on spots, scars and brown marks. Sticks protect you for a longer time because the dense substance adheres well to the skin.

Which factor?

During the first three days, or the first week if you have sensitive skin, you need to maintain high-factor protection for your skin type. If your complexion is very pale, so that you find it hard to get a tan, begin with factor 60 or higher. If it is pale but less sensitive, factor 40 should be enough. If you have a dark complexion, you might start with factor 30 or even 20.

Thereafter, you can gradually reduce the factor: go down from 60 to 40, then from 40 to 20, then from 20 to 10. When you've got the tan you usually achieve, continue to use protection between factors 20 and 10, depending on your skin type. As you slowly reduce the strength of your protection, just as gradually increase the length of time you spend in the sun: 30 minutes on the first day, avoiding the hottest time, then 10 minutes more each day.

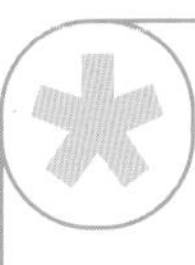

KEY FACTS

* During the first few days, protect your skin with the sun cream most suited to your complexion.

* Gradually reduce the factors until you are down to between 20 and 10.

* Continue to use this level of protection even after you have achieved a deep tan.

27 maximum protection: total sun block

'Total' sun blocks don't, in fact, completely block out the sun's rays, but they do offer the most effective form of protection. They don't prevent you from getting a suntan totally, whether they contain chemical or mineral filters. It just takes longer.

No more white masks

In the past, the application of thick white total sun block in certain places on your body could leave you looking like a rather badly made-up clown. But those days are gone. The new sun blocks are just as invisible as other creams once they have been rubbed in, but they contain much more filter. Mineral filters are the best for this type of product,

●●● DID YOU KNOW?

> Sun blocks are mainly used to protect sensitive areas of skin.

> People who belong to skin type 1 (very fair skin, blond hair and light eyes), post-menopausal women with age spots and pregnant women should all use sun block on their face, neck and shoulders.

> Make sure you keep applying cream regularly while you are in the sun, especially if you go in the water.

> Sun block sticks are easy to apply and handy to carry about.

> Don't just rely on the cream – wear a broad-brimmed hat and T-shirts.

because they leave a protective film on the skin, which greatly reduces the sun's ability to penetrate it.
Sensitive and fragile skin types often require sun block – in which case the mineral-filter creams are even more suitable, because they are far less likely to cause an allergic reaction.

Not quite total

Remember, no sun block can stop all the sun's rays, but they afford enough protection to make the risk of sunburn virtually nil. They won't prevent you from getting a suntan either, but, given the tiny amount of rays they allow through to the skin cells, it will be a long time coming. You'll just have to be patient.
Some brands offer products that add a hi-tech element to their basic effectiveness. For example, there are 'chrono-protection' creams that release their active ingredients gradually on to the skin over a matter of hours so that protection lasts longer.

KEY FACTS

* Total sun block creams provide the most effective protection, although they don't completely block out the sun's rays.

* They are mainly used on sensitive areas: lips, around the eyes, scars and brown patches on the skin.

28

When they first appeared on the market, they seemed like magic: at last you could get a tan without going outside your front door and arrive on the beach looking bronzed on the very first day of the holidays. But the drawbacks soon became apparent.

self-tanning creams: pros and cons

The results depend on your skin

Self-tanning creams contain a tanning agent which helps your skin become golden brown even in the shade, but they don't protect your skin. The tanning is due to the effect on the skin's amino acids of the DHA (dihydroxyacetone) that these creams contain.

Different skin types react differently to them: on some the tan looks com-

DID YOU KNOW?

> Don't confuse self-tanning products with 'tanning accelerators'. These claim to speed up the natural production of melanin by supplying the melanocytes with tyrosine, a vital amino acid.

> Tanning accelerators can be taken in capsule form or applied directly to the skin.

> You can use them whatever your skin type, but you must observe the usual rules of sunbathing.

> However, these products are rated by the US Federal Drug Administration as drugs, and, furthermore, they have not been validated in clinical trials.

pletely natural; on others it can look rather artificial. They can also clog up the pores. Unfortunately, there is no way to know in advance how your skin will react: you simply have to try them out.

Always wash your hands after use

Using a self-tanning product a few days before you go away on holiday means that you don't stick out on the beach among all those tanned bodies. However, remember that your brown skin is not actually tanned, so it will not provide you with any protection from the sun's rays. Some of these products do contain substances used in sun creams, but in self-tanning lotions they are included in such small quantities that they offer no effective defence, particularly during the first few days.

A word of warning: self-tanning creams are extremely potent. You have to be very careful when applying them. Whether you are spreading cream on your face or lotion all over your body, make sure no corner has been forgotten, or it will be very obvious. Also, take care not to leave any finger marks and, above all, wash your hands carefully when you've finished; otherwise you'll have unnaturally orange palms.

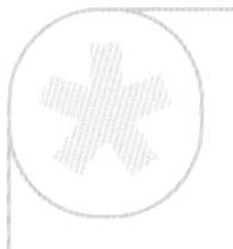

KEY FACTS

* Self-tanning creams tan the skin but provide no sun protection.

* They are effective but be careful when applying them so as not to leave unsightly marks.

* They do, however, provide you with a tan without the dangers of exposure to the sun.

29 regular and generous applications

Buying the right sun cream is not all that matters. You also need to apply it regularly and in sufficient amounts.

Every two hours... To calculate the protection factor of a cream, strict laboratory tests are carried out: 2 mg of the product is applied to each square centimetre of skin. If you apply only half that amount, or fail to reapply it often enough, your factor 25 cream will effectively become a factor 10 cream. As a rough guide, 30 g (1 oz) is usually enough to cover an adult body in each application. Whether you are using cream or oil, apply it half an hour before you leave for the beach, to give it time to start working. Then reapply every two hours.

... or more often After going in the water, dry yourself well with a towel and apply another layer of your skin-protection product. If you are out walking in the countryside, remember to protect all exposed parts of your body. The sun won't ignore you just because you're not sunbathing. The same applies whatever you are doing. Don't think that just because it's cloudy you don't have to wear sun cream while playing a round of golf or tennis for a couple of hours! In fact, if you are active, apply your sun cream more often, as it loses its effectiveness the more your sweat. Also, give your skin a break by wearing a hat, T-shirt and trousers.

DID YOU KNOW?

> Sun cream, no matter how protective it may be, it can't save you from your own excesses.

> Even if you apply a great deal, you shouldn't stay in the sun for more than 30 minutes on the first day of your holiday.

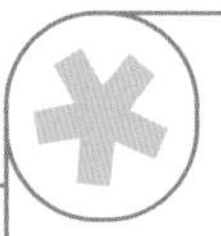

KEY FACTS

* Don't skimp on the amount of sun product you use: too little means less protection.

* Reapply every two hours and each time you leave the water.

30 slop it on all over

Don't forget to apply your sun cream to the less obvious places – lips, hands, neck, shoulders and scalp. They are often the most sensitive parts of the body.

Lips and breasts The lips, with their delicate skin and profusion of blood vessels, are especially vulnerable to sun damage. Use a sun cream stick or special balm to protect them. If you sunbathe topless, apply high factor cream on your breasts. They rarely see the sun and the nipples are especially sensitive, so don't make things more difficult for them than they need to be.

Head and hands Balding men often experience the pain of sunburn because it's easy to forget you have less hair (and therefore less protection) than you did last year. Thorough application of sun cream is the answer. Finally, don't forget your hands. As you use them to apply your cream, you naturally assume they are protected. But while the palms are, the backs may not be. As if to prove the point, that's where the first age spots usually appear (see Tip 18).

DID YOU KNOW?

> Don't spread ordinary sun cream very close to your eyes, because the skin surrounding them is extremely fragile and sensitive.

> There are special sun creams available that are suitable for the area around the eyes.

KEY FACTS

* Certain sensitive areas are often forgotten or poorly protected.
* Pay particular attention to the lips, the area around the eyes, the backs of the hands and the breasts.
* Don't forget that a balding head needs protection as well.

31

think carefully before using a sunbed

Every week a new sunbed salon seems to open up on the high street. They offer you a rapid tan without the need for sunshine and *supposedly* without risk. Be careful, though: don't use them too often, as the rays their lamps produce are even more dangerous than the sun's.

Check those filters!

It's certainly tempting – a few sessions lying under a soft, gently warming light and you've got an instant tan! It seems like a miracle, but the UV rays produced by this equipment are not entirely harmless. Sunbeds in tanning salons emit mainly UVA rays, and have in the past been promoted as 'safe'. However, there's a wide variety of emission spectra and many newer lamps emit a higher amount

●●● DID YOU KNOW?

> Never have two UV sessions in one day. Always allow at least 48 hours between sessions.

> Before lying under the lamps, take off all your make-up, so there's no risk of it causing a mark or allergic reaction. (These rays are much more likely to cause a sensitive response than natural sunlight.) Also, don't wear any perfume or deodorant.

> Finally, don't forget to wear protective glasses, because, even when closed, your eyes are still at considerable risk.

of UVBs to accelerate tanning. The harmful effects of UVA are also increasingly acknowledged. UV lamps should be filtered so that only long-wave rays, those least harmful to the skin, are allowed through, but for this to be the case, the equipment must be well maintained and the filters changed regularly. So it goes without saying that you should only ever use a reputable salon. Avoid salons that allow you to choose your own programme without offering any guidance and go to an establishment where safety guidelines are strictly enforced.

Only now and again

Modern machines deliver UVA at a much greater intensity than the sun, so half an hour on a sunbed gives you more UV exposure than in natural sunlight. Because of this, it's inadvisable to use sunbeds regularly, and they should be avoided altogether if the sun causes you skin problems. You should also bypass these salons if you take drugs that can cause rashes in sunlight (e.g. acne treatments), if you have a history of skin cancer, have lots of moles and freckles, or have very fair skin and don't tan naturally. If none of these apply to you and you feel you must have a tan in the middle of winter, guidelines by British dermatologists suggest no more than two sets of ten thirty-minute sessions per year.

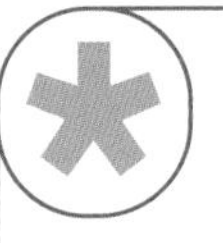

KEY FACTS

* UV lamps in tanning salons produce UVA rays that are much more intense than the sun's.
* These rays increase the risk of skin ageing, wrinkles and cancer.
* Never use a sunbed if you have skin problems, and don't exceed the amount of sessions recommended by dermatologists.

32 relief from psoriasis

Psoriasis is one of the few skin diseases that's helped by exposure to the sun. The flaky patches that are produced become less pronounced and sometimes disappear altogether. This makes sunlight therapy one of the most effective treatments for the illness, but be careful not to overdo it.

A soothing touch

Psoriasis affects 5 to 10 per cent of the population to varying degrees. It's a chronic ailment and is very difficult to treat. However, the sun can have a soothing effect on the white, flaky patches (or plaques) that the disease produces on the elbows, knees, hands, lower back and scalp.

The causes of psoriasis are not completely understood but it's clear that

DID YOU KNOW?

> Water cures are also effective in the treatment of psoriasis. Hydrotherapy is combined with a moderate amount of sunbathing and psychotherapy.

> Many European spas, the Dead Sea in Jordan and the Blue Lagoon in Iceland, offer treatments for psoriasis that can be effective.

there's a psychological element, because the eruptions often occur after emotional shock. Skin cells usually renew themselves about once every four weeks; the old cells come to the surface, die, form loose flakes and fall off. In the case of psoriasis, cell replacement can occur up to four times faster than this, which means the skin can't get rid of the flakes fast enough. These then form a thick layer with inflamed skin underneath.

No more than half an hour

Exposure to sunlight doesn't cure psoriasis, but it does help to relieve it. So do go out in the sunshine as soon as the weather turns fine, but limit your exposure to 30 minutes. Artificial ultraviolet light, sometimes combined with a photosensitizer called psoralen, is also used by dermatologists to treat psoriasis, but don't try to treat it without consulting a doctor or dermatologist first.

> Some people find homeopathy effective. The treatment it provides is precisely adapted to the individual. It takes into account where the flaky patches occur, their appearance and the factors that give rise to them. You can't select your own treatment; consult a homeopathic practitioner.

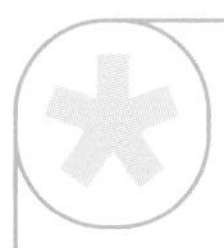

KEY FACTS

* The symptoms of psoriasis are white, flaky patches on the scalp, elbows, knees and face.

* Moderate exposure to the sun (30 minutes per day) relieves the outbreaks, but it isn't a cure and will only help those parts of the body that the sunlight can reach.

33 look after your eyes

The sun can burn your eyes as well as your skin, if you're not careful. The obvious answer is to wear sun-glasses, but make sure they are good ones. Pay particular attention to their UV-filtration and light-absorption percentages.

DID YOU KNOW?

> Plastic lenses are unbreakable but can be scratched. Glass lenses are much more scratch resistant but are heavier and can crack.

> Where there's a good chance of your sunglasses breaking (e.g. when you're playing a sport) opt for plastic or toughened glass.

> If you don't wear sunglasses in bright sunshine, you risk keratitis ('sunburn' of the eye), or tiny burns on the cornea.

Poorly protected

Our eyes are poorly equipped to prevent sun damage. Although our eyelids form a protective barrier, this is only when they're shut, and we can't keep our eyes closed all the time. When they're open, the job of protection goes to the lachrymal film, a watery film produced by the tear glands that spreads over the surface of the eyeball during blinking. However, it's hardly an effective protection against sunlight as it allows most of the UV rays to pass through. Light-coloured eyes are most at risk, as the pigments in the iris provide a less effective screen than those of dark eyes. Whatever the colour of your eyes, however, always wear sunglasses in bright sunshine.

The right shades for every occasion

When it comes to sunglasses, quality counts, but the most expensive may not be the best. Look for an official standard rating. In the UK, the European 'CE' mark and/or the British Standard BSEN 1836: 1997, set performance levels for the amount of UV the glasses let through. Protection factor numbers indicate the degree of protection, with higher numbers affording greater protection. A factor between 1.1 and 2 indicates a cosmetic pair of sunglasses: they reduce light transmission to 30–40 per cent of full sunlight and UVB rays to 3–4 per cent, but have little effect on UVA. General-purpose lenses (shade number 2.5–3.1) reduce light transmission to 8–30 per cent, UVB rays to 1–3 per cent, and UVA rays to 2–8 per cent. Protection factors 3.1– 4.1 reduce UVB to less than 1 per cent and completely block UVA; they are suitable for the tropics or skiing. For driving, use sunglasses with a factor number less than 3, but never wear them at night. Wearing tinted glasses with no UV protection may be harmful as the pupil will react to the reduced light by dilating, so will let through more UV than it would in full daylight.

> In the long term, UV rays can cause cataracts. It is estimated that 10 per cent of cataracts are due to exposure to the sun.

KEY FACTS

* The sun's rays can cause a kind of 'sunburn' of the eye.
* To avoid this, always wear good-quality sunglasses in bright sunshine, but be careful not to wear a very dark pair when driving.

34 hats and sunglasses for the kids

Children are particularly susceptible to the sun, so you have to protect them well. Remember, you're not only protecting them from sunburn, but also ensuring they don't exhaust their reserves of resistance to the sun while they're still young. They'll thank you for it later!

The small and not so small

The younger children are, the more they need to be protected from direct sunlight. Before the age of three, their melanocytes are not fully active, which means their skin's natural protective system isn't yet effective, so they are at a very high risk of sunburn. As children get older, the melanocytes become more efficient, but are only fully functional at

DID YOU KNOW?

> Serious sunburn in a child can increase the risk of skin cancer during adulthood.

> Children under two have difficulty adapting to changes in temperature, increasing the possibility of heatstroke (see Tip 50).

> If your child is red and hot, looks tired and complains of headaches and feeling sick, try to cool them by going indoors, turning on a fan and supplying plenty of drinks. If your child is still unwell, take them to a doctor immediately.

> Give children plenty to drink when they are in the sun, even if they insist they are not thirsty.

about the age of fifteen. So toddlers shouldn't be exposed to full sunshine on the beach after ten o'clock in the morning or before five o'clock in the afternoon in high summer. Ensure that they always wear a T-shirt, even when in the water, and apply generous amounts of sun cream to all exposed parts. Also insist on them wearing a hat and sunglasses. With children older than three, you still need to be extremely careful. Apply plenty of high-factor cream (at least factor 60 if they have a pale complexion; 40 for those with darker skin) and reapply it more often than you would for yourself, particularly if they go in the water. (If they do, don't forget to dry them before reapplying the cream.)

Not all day

Never spend the whole day on the beach or walking in the mountains if you have a young child with you. And remember that a siesta under a sunshade might not fully protect a child from the sun: reflections from the water, sand or rocks could still cause burns (see Tip 35). It's during childhood that too much exposure to the sun has the most serious, long-term consequences and when stocks of sun resistance are most rapidly depleted. Think of your children's future – and protect them.

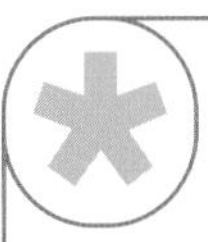

KEY FACTS

* The protective system of a child's skin does not become fully functional until about the age of fifteen.

* No direct, intense sunlight before the age of three, and thereafter maximum protection: cream, sunglasses, hat and T-shirt.

35 beware of the mirror effect

The sun's rays can easily reach you by reflection, so it's possible to burn even when you are in the shade.

Pretty reflections; ugly effects The sun's rays reflect off all bright surfaces. Whether you are at the beach, in the countryside or in the mountains, it's likely that there will be a surface near by that will reflect sunlight on to you. It could be the sand, the surface of a lake or a covering of snow (see Tip 36). When you're on the beach, be careful of the sand, especially if it is a light colour. Its reflections can be so pronounced that you could burn, even if you sit all day in the shade.

Watch out for reflections off water If you spend the day on the water in a boat, there's even more danger of sunburn, because the sparkling surface of the water acts like a mirror. Should your skin also get covered in spray, the tiny droplets of water will increase the power of the rays. So be doubly careful. You are not even safe spending the day in the countryside: be careful of the surfaces of rivers and streams, white rocks and bright stones.

DID YOU KNOW?

> The wind can also be unforgiving.

> It not only dries out the skin, but also has a cooling effect on the skin, which reduces the sun's heat and can easily make you forget how strong it really is.

KEY FACTS

* The sun's rays reflect off all bright surfaces, so they can reach you even when you are sitting in the shade.

* Beware of white sand at the beach, the water around you in a boat and light-coloured rocks in the mountains.

36 just how dangerous is the snow?

In the mountains, the sun's rays are reflected and intensified, so skiers need to be extremely careful.

A gleaming white mirror It would be difficult to find a surface that reflects light better than snow. That is one reason why people tan very quickly in the mountains even in the depths of winter, when the sun is at its weakest. Another reason is the altitude: you are actually closer to the sun, which makes an enormous difference. It has been estimated that UVB rays absorbed at an altitude of 1500 metres (5000 feet) are 20 per cent more powerful than those at sea level.

Take care, even if you can't see the sun Before going out, even on the greyest days, apply a high-factor sun cream and carry a sun block stick that you can use on your lips and other sensitive areas throughout the day. If it's cold, wear a cap with a visor to shield your eyes rather than a woolly hat. If it's fine, protect your arms and neck.

DID YOU KNOW?

> Your eyes are particularly sensitive to rays that are reflected off the snow, so don't forget to always wear a cap and a good pair of sunglasses.

> Remember to choose lenses with extremely high light-absorption and UV-filtration percentages (see Tip 33).

KEY FACTS

* In the mountains in winter, there's a double danger – the snow strongly reflects the sun's rays and the rays are more powerful due to the high altitude.

* So be twice as careful – wear protective cream, cap and sunglasses.

37

use aloe vera gel

Plants can offer very effective protection for your skin. Aloe vera is a Mediterranean plant that contains a gel with extremely beneficial qualities: it is soothing, moisturizing and nourishing. So it is ideal for preparing the skin before you go out in the sun and for relieving it afterwards.

No substitute for sun cream

Aloe vera gel is not a protective cream and it contains no anti-UV filter. However, if it is used in conjunction with your sun cream, it provides several benefits. It contains both vitamin C and beta-carotene, which improve the skin's natural protection against UV rays and it is also very rich in amino acids, including tyrosine, which stimulates the melanocytes.

DID YOU KNOW?

> If you have aloe vera plants at home, you can use their leaves as a treatment for sunburn.

> Cut off the large leaves that grow at the base of the plant and let the yellow sap ooze out. Next, peel off the outer skin and prickles. Then cut the inner part of the leaf into slices and put these on the sunburned areas. Your skin will absorb the gel in about twenty minutes.

> If the burning sensation remains, apply some fresh strips.

Over the last twenty years, numerous clinical studies have confirmed many of the qualities that traditional herbal medicine has attributed to this plant over the centuries. Aloe vera gel cures burns, speeds healing and soothes inflammation. Because it is also easily absorbed by the body's tissues, it provides the skin with deep moisturization and nourishment. Finally, it accelerates the elimination of dead cells, which is particularly useful at the end of the summer to ensure you don't lose your lovely tan too quickly.

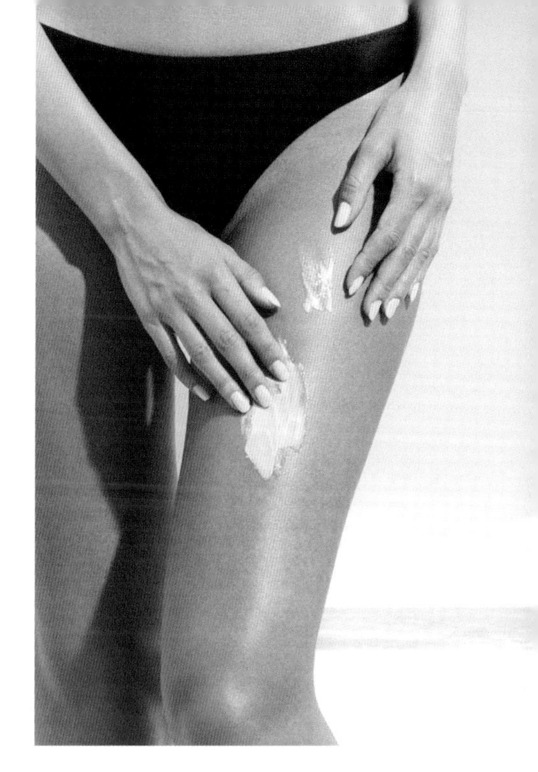

A natural gel

To benefit from all these qualities, just apply aloe vera gel to your skin regularly: before you put on your protective cream and begin sunbathing; and again in the evening when you've finished. The gel is a watery substance obtained from the plant's leaves that is perfectly safe and has never been known to cause allergic reactions. As if all that were not enough, aloe vera gel also accelerates the skin's production of fibroblasts by up to eight times. When the fibroblasts, which produce collagen, are in poor condition, the skin ages and wrinkles appear.

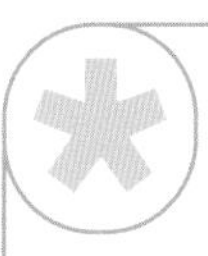

KEY FACTS

* Aloe vera gel helps protect the skin from the sun, even though it doesn't contain any filters.

* It soothes inflammation, improves healing and provides the skin with deep nourishment.

* Put it on before you apply your sun cream and after sunbathing.

38 don't forget shea-butter

Shea-butter (also known as karité butter) is a fatty substance extracted from a nut that has been used for centuries in Africa as a beauty product. Its beneficial properties are due to its wealth of phytosterols, and it is an excellent resource when combating the effects of the sun.

A cosmetic from bygone days

According to certain African legends, the 'butter tree', dating back to the Flood, was given to man by the gods. Nowadays, the shea tree is protected by international forestry agreements, and it is cultivated in several African countries. Its nuts are eaten, and a fat, which was probably the continent's first beauty product, is extracted from them.

DID YOU KNOW?

> Natural shea-butter has a strange, greenish colour and a faintly unpleasant smell. If you don't want to use it in this state, you can choose one of the many products in which it is used a base, but these may be less effective than the pure form.

> All kinds of products using shea-butter are available, from face creams to shampoos, etc. Balms contain the highest concentration of shea-butter.

> Put a balm or pot of natural shea-butter in your holiday suitcase. When you return from your holidays, use a shea-butter body lotion once a week.

Shea-butter is rich in phytosterols, substances found in plants that naturally stimulate the secretion of hormones and the assimilation of vitamins by the skin. So it's a wonderful way to help the skin renew its tissues and improve its suppleness. It also contains essential fatty acids and some proteins, which improve the skin's resistance to the sun and help it repair sun damage.

Rebuilding and hydrating

Shea-butter has a thick, greasy consistency, which some people find a little unpleasant, but it's at its most effective in its natural form. Among other things, it helps to restore moisture to the skin. Exposure to the sun and bathing in the sea both have a severely dehydrating effect.

But remember, shea-butter, like aloe vera gel, is not a UV filter. It should be used in conjunction with skin-protection products, not in place of them. Apply it morning and evening to improve the overall condition of the skin.

KEY FACTS

* Shea-butter is not a solar filter but it improves the skin's tolerance to the sun: it helps the skin renew itself and remain moist and supple.

* You can use either natural shea-butter or products that use it as a base. Of the latter, balms have the highest concentration of shea.

When clouds cover the sun, don't think you're safe from sunburn. Not all clouds filter out UV rays, so continue to protect your skin even when the sun is invisible. And, most importantly, continue to protect your children.

39 don't be fooled by clouds

A thick or thin layer?

A cloud's capacity to filter out the sun depends on its thickness and its density. When you pass through a bank of clouds in an aeroplane just before landing, you see just how different clouds can be: some are dense and as black as coal; others are as flimsy as lace and as white as snow. Although light clouds block most of the sun's infrared rays (those responsible for heat), they stop only a

DID YOU KNOW?

> Take particular care when it's cloudy in the mountains. UV rays are especially strong at high altitudes, so misjudging the thickness of clouds can be more than usually unfortunate up there.

> To avoid risks, continue to protect yourself as if the sun were shining brightly, unless the clouds are really heavy and threatening.

> Follow the same advice when you are busy outdoors, doing DIY or gardening.

small proportion of the ultraviolet rays. Dense, thick clouds are more effective at filtering out UVs, but, of course, they're not particularly common in summer. From the ground, the sun may seem well hidden, but, in reality, it may be lurking behind a very thin layer of cloud.

Don't change your protection routine

Whether you are on the beach, in the country or in the mountains, protect yourself as usual, even when the sky is overcast. Continue to use your sun cream and wear your sunglasses. (The reflection caused by light clouds is very uncomfortable for sensitive eyes.) An outdoor siesta on a cloudy day is particularly deceptive. Because it doesn't feel very warm, you get a false impression of how many UV rays you are absorbing. It's only when you go indoors in the evening that you become aware of how your face is glowing and how burned your skin feels.

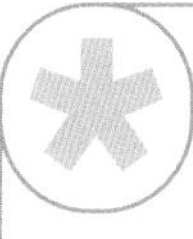

KEY FACTS

* Thin, light clouds filter out infrared rays but not UV rays, which cause sunburn.

* Unless the sky is really dark and threatening, continue to protect your skin as usual.

40 be extra careful with babies

Children under the age of two should never be exposed to any bright, direct sunlight.

Burns and dehydration Direct sunlight on a baby's immature skin can very easily cause burns. Even in the shade, excessive heat can quickly result in dehydration and an unhealthy rise in body temperature, which you won't necessarily notice in time. The cooling mechanism of babies under six months old is not adequately developed to counteract high temperatures.

The solution *Never* take a baby or toddler to the beach. Even sitting under a sunshade, they will still be vulnerable to the reflected rays of the sun, and in a pram the effect will be like being in a greenhouse. Nor should you take a baby camping (in the hot sun, tents can turn into saunas) or for long walks (more than two hours).

DID YOU KNOW?

> If you have to take a baby into strong sunlight, apply a special protective cream. The baby should be well covered in loose clothes made of a natural material, such as cotton or linen, and don't forget a wide-brimmed hat or a cap.

> Keep the baby out of direct sunlight and give them something to drink every 15 minutes.

KEY FACTS

* Babies are as likely to suffer from heatstroke as sunburn.

* They shouldn't be taken to the beach before the age of two, nor taken on long walks in the country or the mountains in hot sun.

case study

Now I always protect my children

«When I was young, I was crazy about sunbathing. The darker my complexion, the better I thought I looked. It was doubtless a way of hiding myself, but I didn't know that at the time. When I had my first baby, I didn't stop to think before taking him to the beach when he was hardly three months old. He was under his sunshade, so I thought everything would be all right. In the evening, when we returned home, I noticed he was very red, hot and exhausted. I was beside myself with worry. Fortunately, it wasn't very serious, but it gave me a nasty shock. After that I was more careful. I didn't take him to the beach again until he was four and then I followed all the advice: protective cream, T-shirt, sunglasses etc. I did the same thing with my next two children. The three of them are now aged from fifteen to eighteen and I'm very pleased with the condition of their skin. They slowly go a lovely golden colour in the sun and never burn. As for me, I'm paying for my carelessness when I was young: I've certainly got more wrinkles than I should have for a woman of my age.»

» **The sun is very demanding:** you have to prepare yourself before going out in it, protect yourself while you are exposed, and take the right precautions afterwards. Only then will you avoid harm.

»» **Basic skin care is vital after sunbathing:** showering, after-sun lotion and facial scrubs, for instance. Plants and seaweed can play their parts, too.

»»» **Some people find the heat especially hard to take,** and need to treat swollen legs or copious sweating.

»»»» Then there are the **emergency measures** for dealing with extreme sunburn and heat stroke.

60
TIPS

41

After you've been sunbathing, jump into the shower to rehydrate your skin and wash away all the sand, salt and dust. Your skin will benefit all the more from the after-sun care you are about to give it … and your tan will glow.

take a shower

Lovely, cleansing water

Perhaps you've been sunbathing on the beach, by the pool or in a field of daisies. Or maybe you have spent the whole day trekking along mountain paths. Your first priority when you get back to your room should be a good shower.
Naturally, the skin dries when it's exposed to the sun. Plus, of course, you

●●● DID YOU KNOW?

> To make your shower even more healthy and refreshing, try the following massage circuit, which will stimulate all your bodily functions, including those of your skin.

> Begin by letting the jet of water play over the sole of your foot. Then, very slowly, with little turning movements, move it up your leg until you reach your hip. Do the same on the other leg.

> Next your hands: massage your palms with the water, then the back of your hands, before moving up your arm to your shoulder. Finally, let the water play a while on your stomach.

have to remember that sea-salt and sand from the beach, chlorine from the swimming pool, perspiration (see Tip 55) and wind burn all contribute towards the drying of the skin too. Before you can even begin to think about some in-depth moisturizing, you must ensure that all these harmful elements that linger are washed away.

A real beauty treatment

You can turn a shower into a real health and beauty session. A good shower stimulates blood circulation, which tends to become sluggish in extreme heat (see Tips 52 and 53), and also soothes your muscles if you've been walking or running around on the beach.

Start with a cool shower to stimulate blood circulation and refresh the skin. Adjust the water temperature so it's comfortably cool rather than cold. Take your time over your shower and make sure that every bit of sand, salt, chlorine and sweat is washed away. Finally, increase the temperature to relax your muscles, if you feel the need.

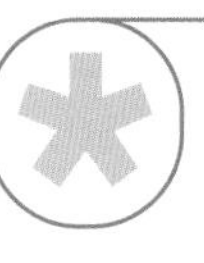

KEY FACTS

* After sunbathing you should take a shower to clean away everything that dries the skin: salt, chlorine, perspiration and so on.

* A cool shower stimulates and accelerates blood circulation, while a hot one relaxes the muscles after exercise.

42 use after-sun lotion

After showering, moisturize your skin. You should do this for at least a month after your holiday. Today's after-sun lotions can also repair cell damage, improve the condition of collagen and increase the skin's resistance to free radicals.

DID YOU KNOW?

> You should continue to use after-sun lotion for several weeks after you return from your holiday.

> It's not merely a matter of repairing sun damage; the skin needs to continue to be nourished and sustained for another month or two.

> Your tan will last longer and your skin will regain its former suppleness and softness more rapidly.

Rehydration is the key

After your shower, make sure you dry yourself well. The kindest way to your skin is to wrap yourself in a large towel or bathrobe that will readily absorb the water and pat yourself dry without rubbing. Now it's time to moisturize. All after-sun lotions contain active ingredients that rehydrate the skin and relieve any sensations of heat and tightness you may feel. The skin's hydrolipidic film, nature's moisturizer, suffers from exposure to sun, water, salt and chlorine, but a high-quality lotion can repair it, so that the skin remains moist for a long time.

Vitamins and minerals

These products also contain substances capable of restoring cellular exchanges, improving resistance to free radicals and even, sometimes, assisting in the repair of DNA in damaged cells. Among these substances are vitamins A, C and E, with the last being the most important. Just as beta-carotene is vital for preparing the skin for sunbathing, vitamin E is essential for repairing it afterwards.

Some of these creams are extremely effective, but you should not treat them as a green light to disregard normal precautions when sunbathing. It is always much better to prevent rather than to cure, so continue to minimize the damage done to your skin by following the advice given earlier in this book: use sunscreen, wear a hat and strictly ration your exposure to the sun.

> There are special after-sun creams for the face that are even more effective than those designed for the body.

KEY FACTS

* After your shower, apply an after-sun lotion all over your body to moisturize your skin and restore the hydrolipidic film, nature's moisturizer.

* After-sun products contain many useful ingredients – notably vitamin E – that can help to repair the skin.

43

If your skin has been more than usually damaged by the sun, if it's very dry, or if fine wrinkles are starting to appear, try some St John's wort oil. Extracted from a plant that has been used medicinally for centuries, this oil can soothe sunburn and heal burns.

St John's wort oil, a natural balm

Scaring demons

In the past, the scent of St John's wort (*Hypericum perforatum*) was reputed to be able to drive away demons. Nowadays, it's often used to treat depression. If you do a lot of sunbathing, you are obviously not at risk of feeling depressed through lack of sunlight, but St John's wort can still help you. The oil is a wonderfully effective means of soothing sunburn and even quickly healing burns. The plant also has antiseptic and other healing qualities: its oil reduces the risk of local infection, soothes inflammation and accelerates healing. It helps the skin to renew itself and regain suppleness.

Home-made medicinal oil

Good-quality, commercially-produced St John's wort oil is widely available, but you can also make it yourself relatively easily. Simply soak 250 g (½ lb) of St John's wort (both stems and flowers) in ½ litre (1 pint) of olive oil mixed with 250 ml (½ pint) of dry white wine for 7 days. Then boil the mixture in an uncovered bain-marie until all the wine has evaporated. Leave to cool and then filter very carefully. Store the red, sweet-smelling oil in dark container with an air-tight lid. You can use this soothing, healing balm in combination with your usual skin-protection measures. The medicinal oil will maintain its potency for two years or more.

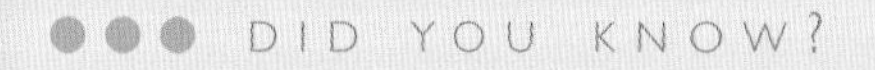

> A word of caution: St John's wort oil contains no sun filter and should not be used instead of sun cream when sunbathing.

> You can use St John's wort oil on your body and face. It's also suitable for use on children's skin.

> St John's wort oil is handy to keep around the home in case of minor injuries such as grazes, cuts and scratches, where it can be used as a natural antiseptic.

> Used in a compress it can help relieve the pain of deep bruising and it makes a good rub for painful joints, varicose veins, muscle strain, arthritis, and rheumatism.

KEY FACTS

* If your skin has suffered a lot of sun damage, or if you have a bad case of sun-burn, try St John's wort oil: it has antiseptic, anti-inflammatory and healing properties.

* You can buy it in health stores and pharmacies, but you can also make good-quality oil yourself.

44

make a fuss of your hair

Hair suffers in the sun: it can be bleached by the hot sun, dried by the wind and the sea, and damaged by chlorine and salt. To keep your hair in good condition during the summer, condition and protect it while you are in the sun. It will be well worth the effort.

Hair bulb, medulla, cortex and cuticle

Hair is made up of many of the same components as the skin in which it is rooted. The hair bulb is embedded in the scalp, from where it draws nutrients from the bloodstream. A hair consists of three layers: the innermost is called the medulla and is almost colourless; the middle layer is the cortex, which contains the melanin that determines

DID YOU KNOW?

> The melanin in hair is not renewable: you cannot give your hair a suntan!

> Dark hair, however, is better protected than light hair, which bleaches in the sun.

> Dyed hair has undergone a process of oxidation that weakens it. Pigments in the hair dye also deteriorate in sunlight and change colour, but there are ranges of products on the market that preserve dyed hair's colour.

hair colour; finally, on the outside, the third layer, the cuticle, is made up of very delicate scales and is kept permanently moist by sebum.
Sunlight cannot cause hair to suffer cell mutation, because it is mostly inert matter (its only living part is the hair bulb), but the sun can dry and weaken it. In addition, salt, sand and wind, among other things, limit the effectiveness of sebum, so the scales on the cuticle are lifted, thus damaging the hair.

Brush, rinse and care for your hair

Begin your hair care by eating well. Hair needs iron and several specific vitamins, especially those belonging to the B group, found in abundance in cereals and dried fruit. Three months before the beginning of summer, you could also start taking food supplements specially formulated to benefit your hair.
While sunbathing, try not to smear your hair with sun oil or cream. Afterwards, brush it gently to remove any sand and dust. If it is covered in salt or chlorine, rinse it in clear water. When you wash it (if not every day, wash it at least every two or three days), use a very gentle shampoo, and rinse thoroughly afterwards. Finally, treat it to a nourishing, protective mask or oil bath.

KEY FACTS

* Nourish your hair – eat plenty of iron and B group vitamins.
* Sun, salt and chlorine all harm the hair's protective sebum.
* Brush your hair regularly, rinse it after sunbathing, use a gentle shampoo and treat it to masks and oil baths.

45

Seaweed contains an amazing variety of minerals and is an excellent treatment for sun-damaged skin. Soak in a relaxing seaweed bath, apply it to affected areas of the skin or take it in the form of dietary supplements – it is highly versatile.

what seaweed can do for you

All the richness of the sea

Seaweed makes the most of its watery habitat, drawing many nutrients from the seawater in which it lives. The seawater contains a wealth of minerals: sodium and iodine, of course, but also magnesium, potassium, calcium, zinc, selenium and iron. Once absorbed by the seaweed, the nutrients become highly concentrated, some by as much as

DID YOU KNOW?

> Numerous cosmetics and food supplements have a base of seaweed.

> Certain varieties of seaweed have specific benefits: spiruline slows the ageing of cells; wrack accelerates the removal of cellular waste; laminaria regenerates the epidermis; and brown seaweed nourishes the hair.

> A word of warning, however: seaweed absorbs pollutants as well as nutrients from seawater. Make sure that the seaweed you use is taken from the open sea.

10,000 times. But even when these minerals are present in seaweed in only tiny quantities, the human body can very easily assimilate them, so seaweed can be very beneficial for sun-damaged skin and hair.

Hormones, proteins and vitamins

Seaweed is also rich in vitamins (particularly beta-carotene, vitamin A), amino acids (which the body converts into proteins) and plant hormones: all known to be excellent for boosting cellular exchange that has been disturbed by too much sunshine. Put seaweed in your bathwater, use it as a mask, apply it as cream or serum – its active ingredients will penetrate deep into your skin. And don't forget that seaweed can also be eaten, as any sushi fan will tell you. If that doesn't appeal, it can be taken as a dietary supplement in the form of capsules or ampoules.

KEY FACTS

* Seaweed contains very high concentrations of minerals, vitamins, amino acids and plant hormones, so it's a good way to boost cellular exchange in sun-damaged skin.

* Seaweed can be used both externally or internally, in the form of dietary supplements.

46 give your skin a good night's sleep

Your night-time skincare treatments continue to work while you sleep, so give them plenty of time to do their job. After all, it's your night off!

Working through the night Sun, holidays and carefree sleep should all go together. If you're sleeping until late in the morning, make sure your skin reaps the benefits. While you rest and recuperate, it never stops working: renewing and repairing damage and eliminating waste. If you have been sunbathing, it has plenty to do. Your skin is busiest around midnight, so try to apply your night-time skincare before then. If you spent a good part of the day in the sun, it's a good idea to have an early night.

Denser and more concentrated Night-time skincare products are usually denser and richer in active ingredients than ones manufactured for daytime use. Essential fatty acids, in particular, are present in more concentrated form in night products, and the same goes for vitamins, fruit acids and plant DNA extracts. Around half an hour before bedtime and before applying your night cream, cleanse your face thoroughly and use a face mask. It will make your night cream even more effective.

DID YOU KNOW?

> Serums contain concentrated active ingredients that are applied directly to the skin. They generally have a pleasant consistency and can be very effective.

KEY FACTS

* At night, your skin renews and repairs itself.
* If it has been exposed to the sun during the day, there is a lot of work for it to do.
* Give it some help by using skincare products at night and by going to bed early.

47 scrub your body

Gentle body scrubs are a good way to prepare your skin for sunbathing, while a series of more thorough ones will regenerate your skin and preserve your tan once you return home.

A thick skin To protect itself from the sun, the skin grows thicker. So, after a few weeks of sunbathing, it has to get rid of more dead cells than usual. Helping it to do this will make your tan glow and last longer. What's more, your skin will breathe better. There's no need to worry that a body scrub will remove some of your tan. In fact, it will do the opposite. Once or twice a week, use an exfoliating gel: it will clean the pores and accelerate the removal of dead cells.

A steam-room scrub This is the best form of body scrub. After relaxing in a moderately hot steam room, the upper layer of your skin will become softer, so treat yourself to an exfoliating massage. As the dead cells fall away, your skin will start to tingle, looking brighter and rejuvenated.

DID YOU KNOW?

> You could also try a Tahitian-style body scrub.

> When you are on the beach, rub yourself with a mixture of fine sand and monoi oil, rinse with fresh water and then apply some more of the oil.

KEY FACTS

* The skin grows thicker when exposed to the sun.
* During the weeks immediately following your holiday, aid the exfoliation of dead cells by applying a skin scrub, either at home or in a steam room.

48 soothing the effects of sunburn

Sunburn, as should be obvious from its name, really is a form of burn and should be treated as such. Fortunately, there are numerous effective, traditional remedies available, which will help relieve the pain and inflammation.

A serious matter

Sunburn should never be taken lightly. However, if your skin is pink and merely feels a little hot, it's not too serious, and your skin will continue to produce melanin. Soothe away any sensation of discomfort with water mixed with vinegar. The proteins in milk also have a mild anti-inflammatory effect on sunburn. Apply compresses soaked in cool

DID YOU KNOW?

> Even if your skin is only slightly burned, don't do any more sunbathing until the inflammation has died down. Restart gradually, always using a protective cream.

> The same applies if you have peeled and your skin is already tanned underneath the peeling layer.

> More serious burns will take several weeks to heal.

> If you have peeled and the new skin is very pale and pink, wait until the damage has healed completely before sunbathing; and when you do, use a total sun block.

skimmed milk to the affected areas or cover them with natural yoghurt. You could also try applying slices of cucumber, potato, or apple (which also help relieve irritation).

If your skin is very red or purple, or if the burning sensation is unbearable and you are experiencing nausea and fever, consult a doctor immediately.

Lily and lavender

If you fall between these two extremes – your skin is red and you find it hard to bear the touch of clothes – there are some other remedies that might prove effective. Besides aloe vera gel (see Tip 37) and St John's wort oil (see Tip 43), you could try white lily oil, which is both soothing and healing, or spread a few drops of pure medicinal lavender essential oil on particularly sensitive areas. Normally, essential oils should not be applied directly to the skin, but lavender oil can be used undiluted. It is soothing and also has anti-inflammatory and healing properties.

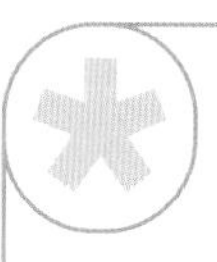

KEY FACTS

* To treat light sunburn, use a little water mixed with vinegar, or poultices of milk or yoghurt.

* For more serious burns, try aloe vera, St John's wort oil or lily oil.

* If you are badly burned and suffering from nausea and fever, consult a doctor immediately.

49 try healing plants

Sunburned skin takes time to renew itself completely, but plants can help the healing process. Many forms of traditional herbal medicine offer natural remedies that help to heal sunburn.

Worth their weight in gold!

Some plants have been used for centuries to help soothe and heal damaged skin, including skin suffering from sunburn damage. The first that deserves a mention is the marigold, a familiar plant in many gardens. These yellow flowers contains allantoine, which encourages the growth of new cells; beta-carotene to improve the quality of the skin's

DID YOU KNOW?

> Arnica (from the daisy family) enjoys a reputation as being the best plant for treating knocks and bruises. Less well known is that it's also effective as a treatment for burns.

> Its flower contains polysaccharides that form a protective layer over a burn while the skin renews itself.

> It can be used as a mother tincture diluted with water, on compresses or as a cream or gel applied to the affected area.

keratin; and mucilages that calm irritation and inflammation. All of these can help improve and accelerate healing. Marigold can be taken in the form of a mother tincture (extract diluted with alcohol and water), and there are many other products also available that have marigold as a base.

Magical mimosa

Another plant with a good reputation for treating burns is Mimosa *tenuiflora*. This prickly shrub is a native of Mexico, where it has been cultivated intensively since the great earthquake that ravaged the country in 1985. It was around this time that the Mexican health authorities rediscovered the extraordinary healing powers of this medicinal plant, called 'tepezcohuite' by the Mexicans. It is also known as the 'skin tree', because its bark immediately reduces the pain of burns and facilitates rapid healing, even after very serious damage. It does not grow naturally in Europe, but healing products that use it as a base are widely available, more often than not marketed under the Mexican name.

KEY FACTS

* When skin is burned, it must be helped to heal quickly and effectively. Some plants have been used for this purpose for centuries.

* Marigold aids the healing process, arnica protects the skin and Mexican mimosa accelerates its renewal.

50 dealing with heatstroke

Heat exhaustion, which can occur after a long period of sunbathing or physical exertion in hot, humid conditions, merits immediate action, especially if the patient is young or elderly. Basic first aid, medicinal plants and homeopathic remedies can all prove effective.

When the temperature rises

When exposed to extreme heat, your body temperature can quickly spiral out of control. Your mouth goes dry, rings form around your eyes, and you feel very ill and nauseous. Act quickly. If heatstroke isn't treated immediately, the increase in temperature can be very harmful. Lay the patient down in the shade on their back, feet slightly raised. Remove their clothing and give them frequent sips of water. Cold compresses can also be applied to the

●●● DID YOU KNOW?

> There are also some homeopathic remedies that can quickly relieve heatstroke.

> Belladonna helps reduce sudden fevers (5CH (dilution factor), 5 granules every 30 minutes). If the victim's face is red and swollen, try Apis 5CH.

> Don't confuse heatstroke with fever. The latter is one of the body's protective mechanisms: it raises body temperature in order to deal with germs and viruses. So, fever can be beneficial, and steps should be taken to

face. Alternatively, place the patient in a cool bath of water (not too cold or it could cause thermic shock) or give them a cool sponge-bath. However, if their temperature approaches 40°C (104°F), and they are obviously unwell, drowsy or delirious, with a hot, red skin and vomiting, seek medical advice immediately.

Willow and meadowsweet

A good plant for reducing high temperature is willow. Its bark contains salicylic acid, the principal active ingredient of aspirin. Meadowsweet contains something similar, and both effectively reduce body temperature in cases of heatstroke. They can be taken by people of all ages as a herbal tea and, in the case of willow, in the form of tablets. These remedies may be effective, but remember that the most important treatment for heat exhaustion is to cool the person as soon as possible and give them frequent drinks of water.

reduce it only when the temperature is dangerously high.

> The increase in temperature caused by heatstroke, however, performs no such positive function and should be treated without delay.

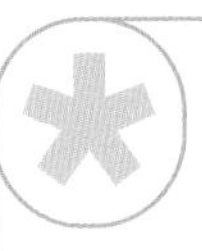

KEY FACTS

* When you spend too long in the sun, your body cannot regulate its internal temperature, which climbs ever higher.

* If this happens, lie down in the shade, put cold compresses on your body or have a cool bath, and drink plenty of water in frequent small sips.

51 act quickly to combat sunstroke

Sunstroke is a combination of sunburn and heatstroke. If you think you have sunstroke, you must take emergency measures immediately: get out of the sun and into the shade, and treat any sunburn. You can also turn to homeopathy for help.

The damage the sun can do

Sunstroke occurs when you have overdosed on the sun. Your skin is burned and your body simply cannot cope with all the heat it has absorbed. The symptoms of sunstroke are therefore those of sunburn and heatstroke combined: a very red, hot skin, an intense sensation of burning, a sudden increase in temperature, which could climb higher than

●●● DID YOU KNOW?

> If children under the age of fifteen have too much sun, the consequences can be very serious: their burns can be severe; they react more violently to heatstroke than adults; and can even suffer convulsions.

> Severe burns in childhood appear to increase the risk of skin cancer later in life.

> Sunstroke is also extremely dangerous for elderly people, because they tend to lose the sensation of thirst and they are less able to regulate their body's temperature than younger adults.

40°C (104°F), headaches and nausea. If your body temperature increases too quickly and remains high for too long, you could permanently damage your nervous system. So you must act fast.

First aid

The essential first-aid measures are the same as those used for heatstroke (see Tip 50), plus treatment of the particular areas affected by sunburn (see Tips 48 and 49). Don't forget to drink plenty of water, as an overheated body quickly becomes dehydrated. If cooling and fluids do not produce a rapid improvement, and especially if there is any drowsiness, confusion or vomiting, seek emergency medical advice immediately.

Homeopathy

As the first aid starts to work, you can supplement it with some homeopathic treatment. For all cases of sunstroke, take 4 granules of Glonoinom 4CH every 15 minutes until you feel better. If your burns are not forming blisters, take 4 granules each of Apis 5CH and Cantharis 5CH every 2 hours. If blisters begin to form, take 4 granules each of Apis and Belladonna 5CH every 2 hours. Again, if your condition does not improve quickly, consult a doctor.

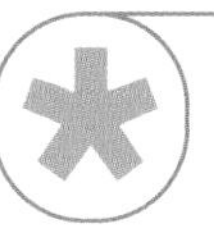

KEY FACTS

* Sunstroke is sunburn combined with severe heat exhaustion.

* Use the same emergency measures as you would for heatstroke and treat your sunburn.

* You can also try a homeopathic remedy to treat your damaged skin and raised temperature.

52 keep an eye on your legs

Many women are aware that their legs tend to swell and feel heavy in summer. This happens because heat disturbs the blood's circulation as it returns to the heart, and also affects the circulation of lymph. If you are particularly prone to these problems, limit the time you spend in the sun.

> To prevent them flaring up, avoid hot, spicy food, alcohol and stimulants, particularly coffee. Drink plenty of other fluids, though, and eat lots of fresh fruit and vegetables.

●●● DID YOU KNOW?

> A long time spent in the sun causes some people to suffer from haemorrhoids.

Blood and lymph

If you spend too long basking in the sun, your legs are liable to swell and feel heavy; sometimes they may even become painful. There are several reasons for this. First, the body keeps its temperature down in hot weather by increasing blood flow to the skin. This increased blood flow ultimately raises pressure in the veins of the leg and produces more tissue fluid, which causes swelling in the legs. The lymph network, which normally drains excess tissue fluid, cannot cope with such an increased load. Lymph, a whitish fluid that circulates upwards from the lower part of the body, is involved in the removal of certain kinds of metabolic waste. Lymph vessels have no pump similar to the heart to help the fluid make this journey against gravity, and when it's very hot, the lymph pools in the legs, especially around the ankles, which makes them swell.

> The homeopathic medicine most often prescribed as a cure for haemorrhoids is *Aesculus hippocastanum* (horse chestnut; see Tip 53). Take four granules of it three times a day.

Put your feet up

If you suffer from circulation problems, don't sit in the sun for hour after hour. This position makes the upward journey of the blood and the lymph even harder. If you do have to sit for any length of time, put your feet up. Do some walking, which will tone up the blood and lymph vessels on the soles of the feet and improve upward circulation. Wearing light clothing and a wide-brimmed hat can also help, and make sure you drink cool water at regular intervals.

KEY FACTS

* Heat dilates the veins and increases blood volume, which causes the legs to swell and become painful.

* The circulation of lymph is also disturbed by heat.

* Don't sit still for long periods in the sun: go for a walk.

53 soothing swollen legs

Some plants, such as horse chestnut, sweet clover and red vine, work wonders for sore, swollen legs, but you need to take them all year round. If you need urgent treatment, however, choose the essential oil of cypress or angelica.

Long live the vine

The vine is an excellent plant for improving venous circulation and red vine has been used for this purpose for thousands of years. Its leaves contract blood vessels, reduce damage to vein walls and generally stimulate the whole network of veins. However, red vine must be taken for several months before it has any significant effect. It cannot

●●● DID YOU KNOW?

> Another plant that is very good for treating problems with veins is the horse chestnut, which has an exceptional ability to narrow blood vessels.

> You can take it internally in the form of capsules or ampoules or apply it directly to affected areas as an ointment.

> Sweet clover can be used to treat varicose veins and haemorrhoids,

provide an instant answer to the circulation problems that occur after too much sunbathing.

Try massage

In such cases, try an essential oil such as cypress tree, which is particularly astringent (it contracts tissues) and strengthens the vein walls. Massage your legs and ankles with a mixture of a base oil (sweet almond, jojoba, avocado, etc.) and cypress essential oil in the proportion of 80:20 respectively. Start by massaging your foot, move slowly up to the ankle, then to the calf and finally to the knee. Work on the sole of the foot in particular, because that's where the vein and lymph networks begin. If the main cause of your problem is the lymphatic system, massage yourself with angelica essential oil. But take care: never use this plant's essential oil before going out in the sun because it reacts to light.

both of which are exacerbated by the heat of the summer months. However, when sweet clover becomes spoiled or contaminated, it can cause internal bleeding. As a result, sweet clover should only be obtained from a reputable supplier.

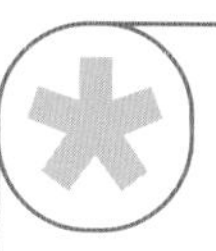

KEY FACTS

* To relieve chronically sore, swollen legs, take red vine, horse chestnut or sweet clover.

* If your problems occur suddenly when you are sunbathing, try massaging your legs with cypress or angelica essential oil.

54 getting rid of excess fluid

If you are not perspiring very much and feel that your legs are swelling in the sunshine, try Java tea and mouse-ear hawkweed.

The problem of water retention If you are susceptible to water retention, the best solution is simply to avoid the heat. But you can't stay inside with the air conditioning turned up high all summer. Fortunately, there are some plants that can help you before you venture outside. Three to four weeks before high summer or the start of your holidays, take mouse-ear hawkweed or Java tea in capsule or liquid-extract form.

Java tea This takes its name from its humid place of origin, and it has particularly good diuretic qualities. It immediately increases the volume of urine and helps remove fluid from the tissues, as well as many toxins. The swelling diminishes, you look thinner and you feel much lighter. Ordinarily, when taken in conventional doses, Java tea is quite safe, but you should avoid it if you have heart or kidney disease, and you should always drink plenty of water when you take it.

●●● DID YOU KNOW?

> Mouse-ear hawkweed is a gentler diuretic than Java tea. Drink it in a herbal tea. Put 50 g (2 oz) of the plant in 1 litre (2 pints) of boiling water and infuse for 10 minutes. Then drink this litre of liquid over the course of the day.

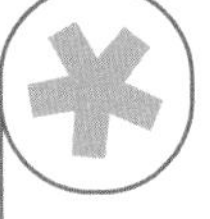

KEY FACTS

* Some plants help alleviate water retention.
* Mouse-ear hawkweed and Java tea both do this effectively.
* Both are safe, unless you suffer from kidney or heart disease, in which case the latter should be avoided.

55 tackling perspiration

Some people react to the heat by sweating a great deal. Here's some advice on how to perspire less.

Too much of a good thing Sweating is a healthy bodily function and is essential for regulating body temperature and removing certain forms of metabolic waste. However, excessive sweating (known as hyperhidrosis) can be a real nuisance. Never prevent your body from sweating by overuse of antiperspirants. These substances close the pores and stop an entirely natural process. Try homeopathy instead. China (chinchona) 9CH reduces perspiration caused by physical exercise, while Pilocarpus 9CH does the same when you sweat for no particular reason.

Avoiding BO Perspiration itself doesn't smell bad: body odour occurs when sweat reacts with bacteria on the skin. To prevent this, good daily hygiene is vital. Also, remove hair from the areas most likely to become odorous and avoid tight-fitting clothes and synthetic fabrics.

DID YOU KNOW?

> Natural products, with a base of essential oils or plants, are available as alternatives to chemical antiperspirants. They stop perspiration from reacting with bacteria and therefore prevent bad odours.

> Good old talcum powder is also effective.

KEY FACTS

* If you sweat heavily, don't overuse antiperspirants and prevent this natural process completely. Try using homeopathic remedies instead.

* BO occurs when sweat reacts with skin bacteria. Some natural products can counteract this.

Traditional Chinese medicine has techniques for treating the effects of sunburn and heatstroke. To relieve the discomfort and pain quickly, try Jin Shin Do. You just have to know the correct acupuncture points.

56 finding the right points

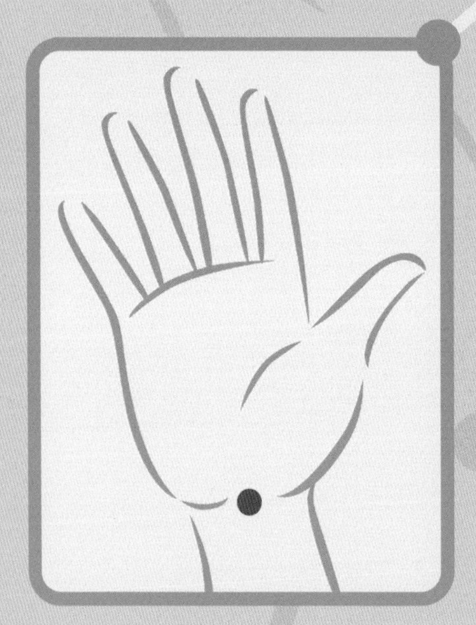

Sunburn: on the wrist

Traditional Chinese medicine is based on the principle of the body's vital energy. If stimulated, your vital energy can help you recover from sunburn. To relieve it, after showering and treating the affected areas with creams or lotions, vigorously massage a point at the base of the inside of the hand, under the palm, in the middle of the wrist. Massage using the soft pad just below the fingertip.

Heatstroke: on the hand

The unpleasant symptoms of heatstroke will disappear more quickly if you work on the following points: first, about an inch below the massage point indicated for sunburn; then on the back of the hand between the knuckles of the index finger and the thumb; and finally, on the wrist in the hollow of the joint.

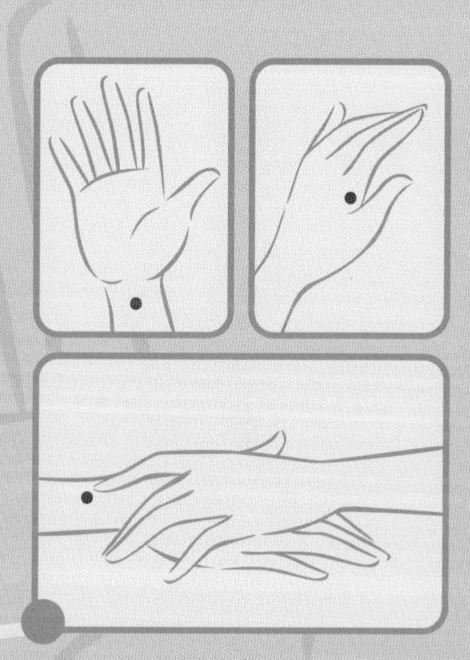

DID YOU KNOW?

> Choose a quiet, cool place in which to give yourself the massage.

> Sit comfortably and breathe gently.

> Press down hard on these points with your thumb or index finger, making small rotating movements.

KEY FACTS

* When properly stimulated, the body's vital energy heals sunburn and heatstroke more rapidly.

* To enable this to happen, vigorously massage specific points on the hands and wrists.

57

A beauty massage is both a pleasurable experience and an effective form of treatment. Learn how to massage your face to tone your skin after you have been sunbathing and reduce the harmful effects of the sun's rays. It takes only 15 quiet, calm minutes in the evening before going to bed.

treat yourself to a beauty massage

A pleasure, whenever you fancy it

You can perform this kind of massage whenever you want. It is soothing and prepares you for sleep. It will also benefit you by relaxing your features, tightening areas of skin that have grown a little loose in the sun and boosting the cellular exchange that stimulates blood circulation. It is ideal for when you have spent long hours laying in the sun.

●●● DID YOU KNOW?

> Massage has an important role to play in all ancient cultural traditions.

> In India, family members massage one another with warm oils.

> In North African countries, skin scrubs and massage form a part of daily hygiene in the hammams (steam rooms).

How to give a self-massage

For an effective massage session, begin by touching the whole of your face very gently and lightly with your fingertips. Then, pressing more heavily, concentrate on your forehead (stretching it up towards the temples), your cheekbones (massaging with small rotating movements), the base of your nose (stretching the skin towards the ears), and your mouth (gently pulling the corners down towards the jaw). Move up to the area around your eyes and gently tap with your fingertips for a few minutes. Finish the session with large, gentle brushing movements across your whole face.
For an even more effective massage, use a home-made oil: mix 2 tablespoonfuls of wheatgerm oil with 5 drops of rosewood essential oil and 5 drops of neroli (orange-flower oil).

> In Far Eastern countries, such as China, Thailand and Japan, massage is thought to improve the circulation of vital energy around the body and thus enhance general health. This includes tackling the consequences of too much sun.

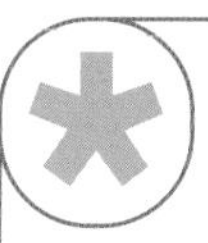

KEY FACTS

* To tone up your skin when it has been damaged by the sun, and to give it back its sparkle, learn the art of self-massage.

* For extra effectiveness, massage yourself with a home-made oil: mix wheatgerm oil with rosewood oil and neroli.

The number of people suffering from sun allergies is increasing. These allergies are not usually serious, but since they make the skin feel unbearably itchy, they can be very unpleasant.

58 dealing with sun allergies

Allergic to what?

A 'sun allergy' is a reaction that occurs when the skin is exposed to sunlight. Its scientific name is polymorphic light eruption, and it has various causes. Usually it's due to a combination of sunshine and another factor, such as drugs, cosmetics or contact with a plant. Research has shown that adverse reactions to the sun usually affect young women under thirty.

●●● DID YOU KNOW?

> It appears that essential fatty acids help protect against allergies and the skin reactions they cause.

> So regularly eat fresh vegetable oils, oily fish and other 'good' fatty acids. Also, try evening-primrose oil or borage oil capsules.

> Pre-sunbathing food supplements seem to help prevent sun allergies, probably because of their vitamin content, which assists all the skin's biological reactions.

A few hours after sun exposure, your skin starts to feel extremely itchy and you might find you are covered with tiny red marks, patches of nettle rash, papules (raised spots) or even blisters, or you could have no visible signs at all. The rash occurs on those areas that have been directly exposed to the sun, such as the face and forearms.

Find the cause

First, try to work out why you are suffering. Are you taking a course of drugs (antibiotics, anti-inflammatory, antifungal, tranquillizers, etc.)? Have you recently changed your cosmetics (face cream or body lotion, for example)? Have you been in contact with any plants, such as while picnicking in the countryside? As with all allergies, the best way of preventing a sun allergy is to avoid all contact with the allergenic substance, so you need to find out what it is.

Once you have discovered the cause, carefully prepare your skin for sunbathing, as recommended earlier in this book, and use a high-factor sun cream. If the rash remains troublesome, your dermatologist may suggest a course of UV light treatment to 'desensitize' the skin.

KEY FACTS

* Polymorphic light eruption is a specific allergy to sunshine, but rashes triggered by sunlight can also be caused by drugs, cosmetics or contact with a plant.

* Symptoms can range from mild to intense itching, as well as a rash.

* Identify the cause and prepare your skin well before sunbathing with a high-factor sun cream.

59

Homeopathic remedies can be effective against sun allergies, relieving the symptoms and preventing recurrences. Consult a homeopathic practitioner, who will be able to choose the most suitable treatment for you.

sun allergies and homeopathy

Diluting the sun

All allergic reactions, including sun allergies, differ from individual to individual. Homeopathy, a branch of medicine that focuses more on the patient and their particular circumstances rather than the illness, therefore can be an effective treatment. One homeopathic remedy in particular appears to suit anyone suffering from a sun allergy: Sol. It is made from a dilution of water that has been

DID YOU KNOW?

> If you have already suffered from a sun allergy, take a dose of Sol or Muriaticum acidum before going on holiday, and then once a week.

> Should the symptoms still occur, also take the remedy that corresponds to your particular condition (see page 121): 4 granules every 2 hours until the symptoms begin to diminish; then 4 granules 3 times a day until they've completely gone.

> If you notice no improvement after 48 hours, consult a homeopathic doctor, who will recommend treatment more precisely matching your case.

exposed to the sun's rays. You could call it 'diluted sunshine'. It is recommended, in highly diluted form (15CH), both as a way of preventing sun allergies and of relieving their symptoms when they occur.

Precisely targeted treatments

Other remedies treat specific symptoms. If you are suffering from pink, swollen rashes that are hot to the touch and itch more when the temperature rises, take Apis mellifica (whole honeybee) 7CH.

If the itching is worse when you get wet, and is exacerbated when the temperature drops, try Urtica urens 5CH.

When your skin has no spots or rashes but still itches intensely, especially at night in the warmth of your bed, use Dolichos pruriens 7CH.

Finally, if you start to itch as soon as you go out in the sun and it worsens when you remove your clothes, Rumex crispus (curly dock) will bring relief.

Another catch-all remedy that can be used whatever your symptoms is Muriaticum acidum (muriatic acid) 7CH.

KEY FACTS

* Homeopathic remedies provide treatments for sun allergies that correspond to the symptoms of the individual.

* Sol is a remedy with a base of sun's rays, which can be used to treat all types of sun allergy.

* Other remedies are chosen according to individual symptoms.

60 dare to be pale

Set a trend and choose not to sunbathe. You will no longer be a victim of the fashion for deeply-bronzed skin.

A lighter tan Thirty or forty years ago, fashion dictated a deep, dark tan. Since then, we have learned much more about the damage the sun can do, and the message is getting through. Nowadays we are generally more careful, and this is reflected in current trends. It's still nice to have a healthy colour, but a dark, bronzed skin is no longer essential. Anyone living in or visiting Australia – a country that has waged an impressively successful campaign against skin cancer – will notice that the young and fashionable rarely have more than a light tan.

Take it a stage further You may not wish to go to the extremes of the past, when only a milky-white complexion was deemed desirable, proving that you were rich enough not to have to toil in the fields to earn a living. These days a light tan is perfectly acceptable and a sign that you are taking care of your body.

DID YOU KNOW?

> **Although our sun is more than five billion years old, it has yet to reach maturity. Scientists say it will exist for at least another five billion years.**

> **Disregard warnings about the power of the sun at your peril. Treat it with respect and stay as safe as possible.**

KEY FACTS

* Fashions have changed and we have now learned how much harm the sun can do us.

* Let's enjoy the sun without feeling we must get a deep-brown suntan all over.

* Instead of sunbathing, cover up and go for a walk in the countryside.

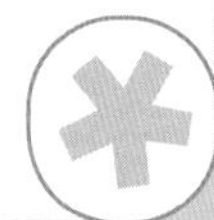

case study

Sun allergies are a flaming nuisance!

« I've always suffered from allergies. That's just the way I am. I've had all sorts, ever since I was a teenager: nettle rash, hay fever, conjunctivitis. I eventually got used to it and always carried antihistamine tablets around with me, just in case. One day I suffered a sun allergy without knowing what it was. I couldn't understand what was going on: there was no rash, no redness and no spots, yet I was scratching myself as if I were crawling with fleas! It was terrible; incredibly itchy. My friends just made fun of me. They didn't believe anything was wrong, because there were no visible signs. I did some research and found out it was a sun allergy. Since then, I've done my best not to go out in the sun too much, I take care about which cosmetics I use and I never do any serious sunbathing. Which is no bad thing, apparently: I'm likely to age less quickly than other people and have fewer wrinkles. So who needs sunbathing? »

useful addresses

» Acupuncture

British Acupuncture Council
63 Jeddo Road
London W12 9HQ
tel: 020 8735 0400
www.acupuncture.org.uk

British Medical Acupuncture Society
12 Marbury House
Higher Whitley, Warrington
Cheshire WA4 4QW.
tel: 01925 730727

Australian Acupuncture and Chinese Medicine Association
PO Box 5142
West End, Queensland 4101
Australia
www.acupuncture.org.au

» Herbal medicine

British Herbal Medicine Association
Sun House, Church Street
Stroud, Gloucester GL5 1JL
tel: 01453 751389

National Institute of Medical Herbalists
56 Longbrook Street
Exeter, Devon EX4 6AH
tel: 01392 426022

» Homeopathy

British Homeopathic Association
Hahnemann House
29 Park Street West
Luton LU1 3BE
tel: 0870 444 3950

The Society of Homeopaths
4a Artizan Road
Northampton NN1 4HU
tel: 01604 621400

Australian Homeopathic Association
PO Box 430, Hastings
Victoria 3915, Australia
www.homeopathyoz.org

» Massage

British Massage Therapy Council
www.bmtc.co.uk

Association of British Massage Therapists
42 Catharine Street
Cambridge CB1 3AW
tel: 01223 240 815

European Institute of Massage
42 Moreton Street
London SW1V 2PB
tel: 020 7931 9862

» Allergies/skin cancer

British Association of Dermatologists
4 Fitzroy Square
London W1T 5HQ
tel: 0207-383-0266
www.bad.org.uk

The British Institute for Allergy and Environmental Therapy
Ffynnonwen, Llangwyryfon
Aberystwyth
Ceredigion SY23 4EY
tel: 01974 241376
www.allergy.org.uk

American Academy of Dermatology
930 E. Woodfield Rd.
Schaumburg, IL 60173-4927
tel: (847) 330-0230
toll-free: (888) 462-3376
www.aad.org

The Cancer Council Australia
Level 5
Medical Foundation Building
92-94 Parramatta Road
Camperdown NSW 2050
tel: (02) 9036 3100
www.cancer.org.au

index

Acne 29, 38–39, 62
Age spots 43, 61, 67
Allergies 44, 69, 118–123
Aloe vera gel 15, 78–79, 101

Beauty spots 36
Blood circulation 37, 89, 108–111, 116
Borage oil 20–21, 118

Calcium 17, 97
Cancer 6, 18–19, 22, 56, 69
Cataract 27
Children 74–75, 84–85, 104–106
Chloasma 42–43
Collagen 19, 32
Connective tissue 18–19, 32

DHA 64
Diets 32, 47

Eczema 44–45
Evening primrose oil 20–21, 118

Fatty acids 20–21, 29–31, 98, 118
Free radicals 18–19, 21–22, 55, 91
Fruit acids 39

Hair 94–95
Heatstroke 57, 74, 84, 104–106, 114–115
Herpes 40–41

Keratin 13, 38

Legs (sore, swollen) 108–112
Lycopene 24–25, 31

Magnesium 17, 97
Moisturizing 14–17
Moles 36

Oedemas 44–45

Perspiration 17, 89, 112–113
Photosensitive (substances) 39, 44
Polymorphic light eruption 27, 118–121
Potassium 17, 97
Pre-sun pills 18–19, 30–31, 33
Protection factor 58–61, 66, 75
Psoriasis 29, 70–71

Rash (erythema) 44
Resistance (reserves of) 12–13, 43, 74–75

St John's wort 92–93, 101
Scars 46, 61
Seaweed 96–97
Selenium 17, 28–29, 31, 97
Self-tanning creams 64–65
Shea-butter 15, 80–81
Singlet oxygen 18, 25, 41
Skin scrubs 15–16, 99, 116
Skin type 12–13, 61–62
Snow 77
Sunburn 6, 73, 84, 100–102, 106, 114–115
Sunstroke 57, 100–101, 106–107

Tanning accelerator 64, 91
Tea 34–35
Total sun block 36–37, 40–43, 46, 62–63

UV (sessions) 68–69

Vitiligo 29, 36–37

acknowledgements

Cover: A. Inden/Zefa; p.8-9: Creasource/Zefa; p.10-11: S. Wilkinson/Iconica; p.13: G&M D. de Lossy/Iconica; p.15: C. Artman/Zefa; p.19: M. Alberstat/Masterfile; p.20: H. Reinhard/Zefa; p.23: Neo Vision/Photonica; A.B./R. Knobloch/Zefa; p.27: M. Tomalty/Masterfile; p.29: M. Alberstat/Masterfile; p.31: J. Burrekl/Masterfile; p.35: A. Bank/Photonica; J. Darell/Getty Images; p.39: S. MacClymont/Getty Images; p.41: S. Wilkinson/Iconica; p.42: G. Contorakes/Masterfile; p.45: C. Sanders/Getty Images; p.49: M. Salmeron/Masterfile; p.50-51: Neo Vision/photonica; p.53: M. Roman/Masterfile; p.55: K. Willardt/Marie-Claire; p.57: P. Cade/Getty Images; p.58: C. von Tuempling/Getty Images; p.61: R. Gomez/Masterfile; p.63: G. Contorakes/Masterfile; p.65: S. Templer/Zefa; p.69: I. Hatz/Zefa; p.71: T. Haus/Marie-Claire; p.72: G&M D. de Lossy/Iconica; p.75: S. Wilkinson/Iconica; p.79: H. Scheibe/Zefa; p.81: Creasource/Zefa; p.83: J.A. Kraulis/Masterfile p.87: H. Scheibe/Zefa; p.89: I. Hatz/Zefa; p.90: W. Packert/Getty Images; p.95: H. Scheibe/Zefa; p.97: E. Hauguel/Marie-Claire; p.101: Star/Zefa; p.103: Neo Vision/Photonica; p. 105: L. Kimmell/ Photonica; p.107: G. Girardot/Marie-Claire; p.108: A. Neunsinger; p.111: I. Garcia/Marie Claire; p.117: G. Chabaneix/Marie-Claire; p. 119: A. Cutting/Photonica; p.121: C. von Tuempling/Iconica.

Illustrations: Christophe Moi: p 92-93, Hélène Lafaix: p 114-115.

60 tips
allergies
60 tips
anti-ageing
60 tips
cellulite
60 tips
detox
60 tips
fertility
60 tips
flat stomach
60 tips
headaches
60 tips
healthy skin

60 tips
sleep
60 tips
slimming
60 tips
stress relief
60 tips
sun protection

Series editor: Marie Borrel

Editors: Delphine Kopff and Caroline Rolland

Graphic design and layout: G & C MOI

Copy preparation: Chloé Chauveau

Final checking: Élodie Ther

Illustrations: Guylaine Moi

Production: Caroline Artémon

Translation: JMS Books LLP

This edition published by Hachette Illustrated UK, Octopus Publishing Group Ltd.,
2–4 Heron Quays, London E14 4JP

English translation by JMS Books LLP (email: moseleystrachan@aol.com)

A CIP catalogue for this book is available from the British Library

ISBN 10: 1-84430-135-4

ISBN 13: 978-1-84430-135-5

Printed by Toppan Printing Co., (HK) Ltd.